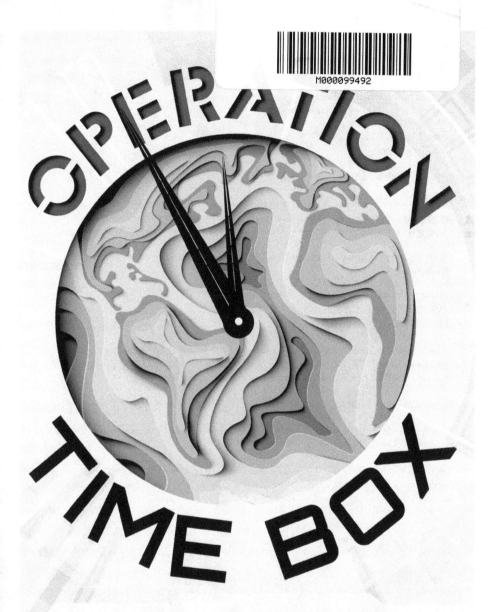

OPERATION TIME BOX

CREATION'S RESCUE MISSION

Gary Wagner

Published by Pen Slinger Publishing
an imprint of Verbal Oxygen Creative Services
P. O. Box 1973
Collegedale, TN 37315
VerbalOxygen.com

Printed in the United States of America

Library of Congress Cataloging-in-Publication Data is on file at the Library of Congress, Washington, D.C.

ISBN 978-1-7320806-0-7 (paperback)
ISBN 978-1-7320806-2-1 (cloth)
ISBN 978-1-7320806-1-4 (ebook)

10 9 8 7 6 5 4 3 2 1

To Deena,

God knew my need and brought you to my side.
You are my friend,
My bag packer,
My adventure partner,
My children's mother, teacher, friend.
Now, my expert editor,
My gift from God.
Always my only love.

CONTENTS

AUTHOR'S FORWARD

There has always been faith, and there has always been science. And for nearly as long, the proponents of either have demanded that the two remain separate. It is becoming increasingly evident that instead of being divided into two separate purviews, the two are more like conjoined twins. Creascience (the study of Creation and science together) is demanding its place in the understanding of all things.

Long ago science had its beginning in the faith community. It was the jealous minds of faith that drove science out. But, thankfully, knowledge could not be suppressed. Now, however, it is science that claims the power to exclude faith from the discussion.

That exclusion is understandable. Self-preservation has always held a dominant place in the affairs of man. Now, that drive to survive has brought intelligence to recognize its need for morality. Science needs faith for its very survival of life.

The re-joining of separated twins is a new experiment, and it doesn't seem to be going well. Without a joint effort, hope is extinguished.

Both realms warn of a climactic end of life-as-we-know-it. Could it be that bringing them back together will produce the only hope for the survival of the masses? Could ending the feud bring science and faith together to facilitate all to receive the gift of eternity? It is, after all, the goal of both sides. And both claim the deadline is near.

I am not a scientist, but I have a deep respect for science and scientists. My two years as a biology major do not make me an expert in any scientific field. My studies are part of what has led me to look at origins with a biblical view. Science, and pseudo-science, are filled with far too many theories that are unproven, unprovable, and even proven wrong, but still demanding to be the only possibilities considered. Intimidation, money, and strong-arm politics have driven origins research and continue to do so. These forces are just as wrong as were the church tactics of previous centuries that sought to limit the power of pure science to maintain their ideas.

The science discussed in this work is not intended to be of a level to satisfy or convince the scientist. It is to give a lay reader a chance to hear some evidence in a form they can understand.

Science, like all other disciplines or occupations, is neither all good nor all bad. Statements in this book are not intended to paint science or scientists with a broad brush of good or evil. Science, in general is good, but not all of what passes for science is science. True science and true Bible-based knowledge will move together hand-in-hand. They both come from the same Source. If one seems irreconcilable with the other, it is because our understanding of one or both is wrong.

In recent years, these topics have been widely debated. Some debates extend for centuries. I am open to comments that seek to justify both science and the Bible. God is. He is Creator. Science is part of what He created. Rightly understood, the two will agree. I have attempted to prayerfully research the Bible for clues about Creation and add to them portions of science that support them. The way these pages describe Creation and science undoubtedly include some errors. My prayer is that the ultimate truth of God's Word will shine through despite my human inadequacy.

INTRODUCTION

The study of Creation reminds us of at least five important issues. First, God made the earth.[1] He made this place special. He made it for you and me. We did not develop over millions of years through natural selection.

Second, Creation shows the character of God. By looking at the things God made, we can understand Him better, even the invisible things that aren't mentioned in the Bible.[2]

Third, God made us in His image.[3] We can learn about the original character God designed in us. We lost that character because of sin. He wants to return us to His image, to be in His likeness. If we allow it, God will recreate His image in us.

Fourth, the worship of God as Creator is among the most important issues shortly before Jesus returns. We are told to worship Him as Creator.[4] We need to understand what that means for our daily lives.

Finally, Creation demonstrates the great love of God and His amazing design for you and for your eternity. If we know more about Creation, then we will understand more about our place in His eternal plan.

God's purposes for creating our earth and us must be many and varied. High among them seems to be that He wanted someone to love, and who would love Him. He wanted a meaningful relationship with beings He could reason with. He made us in His own image to be higher than angels. In terms of today, we were made to show the universe who He is. We are His selfies, designed to share with everyone on His "friends" list, and everyone He wants to invite to be His friends. His desire is that when others see us, they will know something about Him, and they will want to be His friend, too.

Chapter 1

Creation, Day 1

SAVE THE WORLD BEFORE YOU BUILD IT

As with all raging beasts, this one had to be contained. He had proven impossible to restrain in his natural habitat. Already he had provoked the heavenly beings. No one was safe in his presence. All efforts to deal with him had failed. There had to be a way to prevent him from roaming to other places where he would continue to wreak havoc.

A rescue plan had to be developed that would save not only his primary target—the inhabitants of Earth—but also the rest of the universe. These restrictions had to be firm. But the plan also had to be loose enough to allow the beast to show his true colors so that no one would ever be besieged by him again.

And so, a cage had to be constructed. We'll call it a box. It would be a box within which the beast could still roam. If not, some might say he received unfair treatment. There would be others who would be confined within the box as well, but not because they were dangerous. They needed to grow and learn to trust their Master. They would be overcome by the beast if they didn't choose to keep themselves close to the Master and separate from the beast.

There had never been a box like this. It was invented specifically for this rescue mission. The box is called "time." And the beast? You've already met him. But he will come into this narrative later.

So, where shall we start?

"In the beginning…"

God's realm is eternity. Before Creation Day 1, time didn't exist. Although eternity existed, it is not measured in seconds, minutes, hours, days, weeks, and years. How long is eternity? That question is unanswerable. It is also flawed. Eternity is just there. It is not measured at all. Eternity has no measure.

However, there is a significant reason why we needed time. Time began on Day 1. "In the beginning…"[1] Notice that word: beginning. It is the first measure of time. Beginning denotes time—it marks a time when something starts. The Creation account not only tells us about the beginning of this earth, but the beginning of time itself. The things that define you and me do not limit God. He does not only live in the present. He lives in all places and all times. God is now present in your life and mine. He is also now present in the time when you were born, and in the time when you will die. Theologians call this "Omnipresence." It is an attribute belonging uniquely to God.

You and I may sometimes have difficulty remembering the details of our past, especially as we grow older. But it is even more difficult for us to remember the details of our future! This is no problem for God. To Him, all time is now.

Time or space do not confine God. They are essential to see what He has done to be able to bring us into a lasting relationship with Him.

TIME PLACES A CONTROL ON EVIL

Often, we say, "I don't have enough time. Time is going too fast." We may wish we could control time. But the reality is, time exists for our benefit. It draws us closer and closer to the end of sin in the universe. Even though we may sometimes think time is a curse, it is among our choicest gifts.

Before God created the first speck of cosmic dust for this earth, He knew humans would sin. He knew there would be sin in the world. This will come into play with what He did and why He did it on every day of Creation. His goal was to limit the reign of sin and its duration.

God did not want rebellion to be eternal. How could He restrict it? Through the use of time sin would be allowed to exist. Time became a box into which sin could be placed, and the box would be destroyed when it is appropriate to do so.

From the dawn of Creation, God planned to bring an end to sin. Every second you live is one second closer to the completion of God's plan to destroy sin and its impact on you and me. Every second we live is God's promise to us that sin will not last forever. Time is in our favor.

ALL THINGS ARE NEW

We will learn throughout the Creation narrative there are elements, actions, or ideas that never existed in our universe before earth's creation. To draw attention to these aspects and to help us to recognize their importance, they are repeated over and over throughout the first two chapters of Genesis.

From the beginning of Day 1, God began teaching us about time. Each day He said, "the evening and the morning were the first day," "the evening and the morning were the second day," and so on through the entire week. He exhibited the measuring of time. Why was it necessary for Him to describe time? It's because time never existed before. This is a principle we will see all through the Creation narrative.

Will time be present in heaven? We know there will be Sabbaths,[2] but there will no longer be days and nights. The Bible tells us that Jesus will be our Light,[3] and that there will be no night there.[4] If there is no darkness, then there will be no evenings and mornings. So, there will be no time. There will no longer be any need for it. When Jesus returns to take us to heaven, time ceases to exist.[5]

Today, because we still live within the limits of time, we can look forward to the moment when God will wipe away all sin. Sin will not continue forever. It will end. God gave us this promise in the beginning when He created time. There is more to learn about time. We will examine it further on Day 7.

"In the beginning…" marks the dawn of the earth and not of the universe. This is often verified in the text by reminding us that "the evening and the morning were the first day," "…second day," "…third day," etc. The Bible is explicit.

"In the beginning, God created the heaven and the earth." God made our world out of nothing. He fashioned us out of nothing. Yes, He formed Adam out of the dust, but He created the dust at the beginning of the week. God wasn't dependent upon elements that had come before. He wasn't dependent on things anybody else had already created.[6]

BIG BANG THEORY

The best idea man has come up with about the beginning of the universe is the Big Bang theory. It purports that some unexplained power pressed all existing matter into a tiny molecular pinhead of space. The amassed energy caused the compressed matter to expand, scattering it throughout the width and breadth of today's universe. This expansion created planets, suns, and all manner of other flying objects. These spheres miraculously ordered themselves into the orbital configurations of solar systems and galaxies.

It has not been explained where the matter came from before it was compressed, or where space came from. With this and all other theories, there is always pre-existing matter. No plausible explanation other than an intelligent being has been given to explain the origin of primary matter—the stuff which came first. All possible reasons require faith to some degree. You must ask the question, "Am I willing to believe in the spontaneous generation of rocks and elements out of nothing, or in an omnipotent Creator with a plan?"

The Second Law of Thermodynamics says that disorder (a.k.a. entropy) stays the same or increases over time.[7] Still, scientists report the Big Bang formed all the extremely orderly solar systems and galaxies. The word "explosion" gives the impression of massive and immediate disorder from something that was much more orderly. That's why this event is not reported to be an explosion, but rather an "expansion or inflation" of all matter in the universe. This expansion is said to have been at a rate many times the speed of light.[8] No matter what term is used to describe it, this rapid growth shows disorder instead of order. The idea of universal expansion occurring at the rates theorized, while maintaining the systemic order we see in space, is difficult to imagine.

Another reason to avoid the word explosion is that explosions do not create. They destroy. Every atomic and nuclear bomb testing has utilized the same elements credited to the Big Bang. An atom, or the nucleus of an atom, is bombarded with vast amounts of energy until it blows apart.

Since 1945, when the United States used atom bombs in Nagasaki and Hiroshima, there have been many of these great explosions on earth. It is true that none of the blasts approaches the reported power of the Big Bang. It is also true that the only result from every one of the explosions was destruction. Not a single planet or moon has been created, let alone a solar system or galaxy. Absolutely nothing has ever been created by these man-made mini-bangs.

According to some scientists, there may even be more than one universe. Some parts of the original expansion or inflation could have expanded at different rates of speed. If so, this puts them out of sync with our universe. They could have become their own universes with some of their own laws of physics.[9]

Science can represent these theories with formulas and charts, but they are no more verifiable than Creation. Each begins with assumptions that may or may not be accurate.

WHO SETS THE STARS IN PLACE?

The Bible references the stars of our galaxy, with many named in Scripture. "He (God) counts the number of the stars; He calls them all by name."[10] Among the constellations and stars mentioned are the Bear, Orion, the Pleiades, Mazzaroth, the Great Bear with its cubs, the Serpent, and the Twin Brothers.[11]

When God created our world out of nothing, He also established the rules of our world. He made the laws of nature and put them into effect. It was He "who kept the sea inside its boundaries as it burst from the womb."[12]

When God created the oceans, He also set the limits of the waters, so they would not overflow and envelope the earth. In the same way, He created all the laws of our natural world.

FORMLESS AND EMPTY

The Bible tells us what our world was like on the day God began His work of Creation.

"Now the earth was formless and empty, darkness was over the surface of the deep, and the Spirit of God was hovering over the waters."[13]

What does "formless and empty" mean? What did it look like? A tennis ball, for example, has a definite form. It's round or spherical. When the Bible says the earth was "formless," that must mean it was not round like a tennis ball! Nor was it in any other shape easily described. We've traditionally pictured our earth, when God began His creative work, as being a sphere that He hung in space. But if the earth was "formless and empty," then it wasn't a symmetrical ball. There was also some aspect that included emptiness.

There are so many questions. The verse tells us, "the Spirit of God was hovering over the waters."[14] What are the waters of the deep? Are they any different than just "waters," or is the deep the same as the waters? Day 2 will give us more information to determine this.

We do know because of this and later descriptions that the Earth at this point was a place unlike what we are used to today. Humankind has never seen a place like this. God began His work by creating all the needed elements: water, dirt, gems, minerals, metals, gases, etc. God knew what the foundations of the world needed to be to use as building blocks. He created the foundations by first creating all the necessary components to bring this globe into existence. He cast them into a carefully chosen, empty place in space.

SPACE IS BIG. SPACE IS DARK. SPACE IS COLD.

What do we know about space? It is dark. It is expanding in its dimensions. It is cold. How cold is it? Science tells us it is -455 degrees Fahrenheit[15] unless there is a nearby sun. But on Day 1, our sun did not exist.

What happens when you throw water into sub-zero space? The water freezes in mid-air! Picture what happened to the minerals, dirt, and water tossed into this place in space. These elements gathered together and became one great clump of ice, dirt, and minerals, without form. They were not packed tightly together in a ball. The earth did not have a definable shape. Instead, the ice and gravity held it all together.

Gravity had already been created. Whenever you have a mass of anything, it generates its own gravity. The laws of thermodynamics were also already in place on Day 1. This world of ours was "without form and void," with space between the connected and unconnected ice crystals. Some of these crystals may even have floated into outer space.

The lack of a real shape comes from all the elements being put in space and talking about it before the forces of the laws of nature bear on them and turn them into a sphere. This is the principle behind hydrostatic equilibrium. "Stars, planets, and moons can be made of gas, ice or rock. Get enough mass in one area, and it's going to pull all that stuff into a roughly spherical shape."[16]

There needed to be a way to melt all the ice and some other elements, allowing the gravitational pull to bring it to the center quickly. Then, the process called hydrostatic equilibrium would allow for the earth to take spherical shape quickly and develop a smooth rotation.

Job describes the ice. "Does the rain have a father? Who fathers the drops of dew? From whose womb comes the ice? Who gives birth to the frost from the heavens when the waters become hard as stone, when the surface of the deep is frozen?"[17]

Remember we have already noticed the same terminology, "darkness was over the surface of the deep."[18] Through the dark part of Day 1, earth was inhospitable to any form of life that we know today. It was incredibly frigid, with no atmosphere and no useable form of nourishment. But that was about to change.

FROM DARKNESS TO LIGHT

Light was created. "Then God said, 'Let there be light,' and there was light. And God saw the light, that it was good; and God divided the light from the darkness. God called the light Day, and the darkness He called Night. So the evening and the morning were the first day."[19]

The fact that an evening and morning already existed means the earth was rotating. This rotation is vital for several reasons that we will examine later. The light and darkness were just like what we experience today. The earth rotates, allowing every part of it to bask in the source of light.

But what was the source of the light on Day 1? Where did the light come from? How could there simply be light with no source? It wasn't the sun. The sun wasn't created until Day 4. This light had to be something that was available to shine on the earth. Then that source was altered or removed when the sun was added to the picture.

The Bible tells us where this light came from. "In the beginning the Word already existed. The Word was with God, and the Word was God. He existed at the beginning with God. God created everything through him, and nothing was created except through him. The Word gave life to everything that was created, and his life brought light to everyone. The light shines in the darkness, and the darkness can never extinguish it."[20]

Jesus is the Word that John is talking about in these verses. "So the Word became human and made his home among us. He was full of unfailing love and faithfulness. And we have seen his glory, the glory of the Father's one and only Son."[21] Jesus was the Light.

TYPE AND ANTITYPE

A standard tool used in the Bible to make comparisons is of type and antitype. The type is mentioned with some characteristic(s) and points forward to something or someone else. That which is pointed forward to is the antitype. It is the truth to which the symbol points. We can learn about the antitype by examining some of the characteristics of the type. This tool is frequently used in the Creation narrative.

Revelation tells us that in heaven there is no darkness because Jesus will always be present. The sun and moon are types of Jesus, "the Light of the world."[22] When Jesus claims His rightful position in Heaven, the types will not be needed. They may or may not go away. Jesus, the Light is the antitype. Speaking of the New Jerusalem, John wrote, "I did not see a temple in the city, because the Lord God Almighty and the Lamb are its temple. The city does not need the sun or the moon to shine on it, for the glory of God gives it light, and the Lamb is its lamp. On no day will its gates ever be shut, for there will be no night there."[23]

Lamb is another type of Jesus. In John 1:29, we read, "Jesus is the Lamb of God who takes away the sin of the world!" Even the Old Testament identifies the light and describes Him as the Sun. "But for you who fear my name, the Sun of Righteousness will rise with healing in his wings."[24]

This verse is a description of character. The sun has no righteousness. Inanimate objects don't have character and are never described as righteous. The passage shows that this "sun" is not our present-day sun. This is one of only two places in Scripture where it talks about Jesus being the S-U-N. Jesus was the S-O-N; not just the sun of light, but also the Sun of Righteousness. It tells us that He would arise. The use of specific words here (e.g., sun, arise) may well have been intended to provide a double entendre until the explanation could be discerned after the first coming of Jesus.

First, the Book of Job speaks in language that can have double meaning. "By the breath (or voice?) of God ice is given, and the broad waters are frozen.[25] Then in the Psalms where we are told God "sends out His word (or Word?) and melts them; He causes His wind to blow and the waters flow."[26]

According to His plan, Jesus, the world's Creator, came to save those who would overcome the naysayers and accept Him as Lord and Savior. We will see in the Creation narrative more evidence of God's plan to save each one of us. Jesus was Light at the beginning. We will still have His Light at the end of the world when He returns to take us to heaven.

Jesus' presence is so bright that it will destroy all the unrighteous who will see Him when He returns at the Second Coming. "Then the lawless one will be revealed, whom the Lord will consume with the breath of His mouth and destroy with the brightness of His coming."[27]

There is more to learn from Jesus' statement that while He was here, He would be the Light of the world.[28] But Jesus is no longer here in person. He no longer walks the earth. His Holy Spirit is here working among us and in us.[29]

SO WHERE IS HIS LIGHT?

The Apostle Paul tells us, "you may become blameless and pure, children of God without fault in a warped and crooked generation.

Then you will shine among them like stars in the sky as you hold firmly to the word of life."[30] Jesus has made us to be the lights to shine in His place. He intends that we shine with His goodness. To do this, we must hold fast to the Word, which is Jesus. He made it even more clear by saying, "You are the light of the world. A city that is set on a hill cannot be hidden."[31]

When you accept Christ, you become the light to the world. He made you in His own image (see Day 6). You are the light that is intended to shine on the world around you so that everyone can see God in His great glory and recognize that they, too, are made in His image. He has an eternal plan for every person.

The presence of a real hundred-dollar bill reveals the imperfections of a counterfeit. Some want to be viewed as a follower, but without a genuine commitment to Jesus. A faithful follower of God exposes the hypocrisy of a non-follower. This is one reason people sometimes dislike and distrust Christians. Their inner sense recognizes the bright light of the grace-filled Christian reveals their evil or their sin. This occurs even if the Christian makes no effort to expose the wrong in the non-Christian's life. God made us lights to shine in the world around us. When the light of Jesus is shining through us, it cannot be hidden. It is like a shining city on a hill.

God went to considerable trouble to create this world. Even before He made the world, God loved us and chose us in Christ to be holy and without fault in His eyes. "God decided in advance to adopt us into his own family by bringing us to himself through Jesus Christ. This is what he wanted to do, and it gave him great pleasure."[32]

WE CAN COUNT ON GOD

There was another purpose for having Jesus as the Light of Days 1-3. Creation is filled with object lessons to teach us about the plan of salvation. The evening (dark) part of earth's rotation of Days 1-3 was dependably followed by the morning (light) part of each day. The darkness of life is dependably followed by Jesus the Light. Whenever we experience sin, pain, and misery, we know Jesus is near to provide for us and to save us. We can count on it, as much as we are aware that morning always comes after night.

Everything about Creation shows why God made this world for us. He created this earth because of His great love for us and His great desire to be with us. He still loves us and wants to be with us.

No wonder so many influential people and organizations in our world today don't want us to consider the truth of Creation. They don't want us to recognize who God is, and what He placed us here to be and do. They don't want us to realize that His power is available to us and that through Him, we can have the hope of life eternal.

Chapter 2

Creation, Day 2

SETTING THE FOUNDATIONS IN ORDER

WHAT MAKES A DAY?

Consider this basic information. When Creation Day 1 was over, what happened next? Sounds like a trick question, doesn't it? But there's an easy, obvious answer: Day 2 started! That is more important than it may seem. To understand the timing of the creation of our world, we need to know what makes a day. This issue is often discussed, requiring the explanation to be a bit belabored.

From evening to evening, the length of a day is 24 hours. What determined the length of Day 1? The sun and the moon? No, the sun and moon didn't exist on Day 1 or Day 2. Yet the Bible account insists, "the evening and the morning were the first day;" "the evening and the morning were the second day."[1] This is repeated each day of Creation to drive the point home. It's as if God is saying, "Look, My children. This will be an issue in the future. I am repeating this every day, so you will know, without a doubt, that the earth was created in six, twenty-four-hour days."

The length of a day is determined by how long it takes for the earth to rotate on its axis a full 360 degrees. This rotation causes a day on earth to last 23.934 hours (or 23 hours and 56 minutes).[2] We call it twenty-four hours. As the earth turns on its axis, each region of the earth enters a time of dark and a time of light. Until the earth returns to its original position, that "day" continues. Since the earth was rotating during Creation, there was an evening and a morning the same as we have now. The duration of each day of Creation was the same as the length of each day now. It was not a billion years or even a thousand years in length. There was an evening and a morning, and that was a day. This helps us know that Day 2 began at the end of Day 1, and that Day 1 lasted 24 hours.

Jesus was the only light during the first three days of Creation. As the earth rotated away from the place where He was, the light faded. When the earth had completed its full 360-degree rotation, it returned to where He was, and the light returned, bringing the dawn of the second day. Why does this matter? Time did not exist before Day 1. As we learned in Chapter 1, the eternity of God's existence before Day 1 had no time. Eternity is not measured in seconds, minutes, hours, days, weeks, or years. Time was created at the beginning of the Creation week.

This is important to understand because some try to combine the Bible description of the Creation week with the theory of evolution. To do that requires millions and even billions of years to bring us to our present state of being. One way this is done is to use the "Gap Theory."[3]

There are many different versions of this theory. They say the materials for the earth were put here millions of years earlier and then worked on beginning with Day 1 or 2. A huge problem comes when proponents place billions of fossil animals during this gap period. This would allow for death, bloodshed, disease, and suffering before Adam's sin.[4] The word "beginning" and defining Day 1 as the evening and the morning dispel this idea.

When God began Creation week, He not only created light, atmosphere, dry land, and so on. He also created time, as we have already noticed. The substance of the earth had not been present for billions of years—God began creating at the beginning of Day 1, which was a 24-hour day. He continued to create on Day 2, another 24-hour day.

God could have spoken the world into existence in one split second. Or He could have taken a month, a year, or a billion years to create it. But He chose to take six days, consisting of evenings and mornings, as He created time.

Day 2 started with a new evening. Was there light? Yes, there was light. But as the earth turned away from Jesus, there was darkness over half the globe. In the same way, we now have night when our side of the earth rotates away from the sun.

This tells us something about the creative power of God. Why didn't He describe each day of Creation by saying, "the morning and the evening were the first day," "the morning and the evening were the second day"? Instead, He began with darkness. Everything in Creation teaches us three lessons: about God, ourselves, and God's relationship with us. He has the power to bring light out of darkness. When He works His creative miracle in our lives, the same thing happens. God knew that sin would enter the world through man and woman. Sin is depicted by darkness. Every day He calls us "out of darkness into his marvelous light."[5]

As a member of the Trinity of God,[6] Jesus had the ability to be everywhere at once. During Creation, He must have stayed in one place. He did not use His omnipresence during Creation week, or there would have been light everywhere for the entire 24 hours. He may have chosen to temporarily give up that power during the Creation of this earth, perhaps as a foreshadowing of His sacrifice for sin which was yet to come.

SEPARATING SKY FROM EARTH

On Day 2, God said, "'Let there be a space between the waters, to separate the waters of the heavens from the waters of the earth.' And that is what happened. God made this space to separate the waters of the earth from the waters of the heavens. God called the space 'sky.'"[7]

SPACE ICE

We have the ability today to see, to some degree, what the earth was like at the dawn of Day 2. We can look at space ice. "Space ice is made from a mixture of methanol and water. It expands under pressure and shrinks when heated."[8]

Comet 67P is one example of space ice. In 2014, the European Space Agency explored this comet with the use of a space module to try to discover where the comet came from and what it was made of. The shape of this comet is what we could call "without form." It's not a shape that you can categorize, except that it looks like a giant rubber duck.[9]

Space.com reports the comet's length is about three miles, and it is composed primarily of ice.[10] It is interesting to note these are the same materials God threw into this place in space to form our world as it is now. Perhaps 67P's make-up will show that it is a piece of ice that escaped when the earth was being made, and it flew off into outer space only to keep returning to us in its space orbit. Scientists, however, believe it comes from a region of space "beyond Neptune's orbit."[10a]

The space module has "directly detected a crucial amino acid and a rich selection of organic molecules in the dusty atmosphere of a comet, further bolstering the hypothesis that these icy objects delivered some of life's ingredients to Earth."[11] This belief may come from their preconception that life came to earth from space. It is unlikely they have considered a possibility that it came from here and began its regular journey to Neptune and back.

Along with Earth's ice, many other elements mixed in with the dirt, similar to the 67P Comet, just waiting to be formed by the Master's hand. While some of the raw materials may have been tightly formed, perhaps others scattered out into what we now call our atmosphere.

MELTING THE ICE

Science tells us "Earth is thought to have arisen from a cloud of gas and dust in space. Solid particles, called 'planetesimals' condensed out of the cloud. They're thought to have stuck together and created the early Earth. Bombarding planetesimals heated Earth to a molten state."[12]

This takes as much faith to believe as does Creation. First, the planet is "thought" to have arisen from a space cloud of space dust that is "thought" to have stuck together. This must also mean that heating the Earth to a molten state is also "thought" to have been caused by bombarding planetesimals. Of course, I can't prove the existence of Creator God, either. Still the evidence available from the entire narrative is at least equally compelling.

The ice had certainly begun melting during the "morning" segment of Day 1. At the beginning of Day 2, all around the world the ice continued to melt. As it did so, other elements of the earth were drawn together. Water began to wash down toward the center of gravity. Perhaps other elements also melted, as they will after the final judgment and destruction of the wicked. "But the day of the Lord will come as a thief in the night, in which the heavens will pass away with a great noise, and the elements will melt with fervent heat; both the earth and the works that are in it will be burned up."[13] This is the same heat that Jesus brought into action on Day 1 when He melted the ice and other substances He cast into this place in space.

Jesus, the light source, has the power to melt every element contained in the earth. The coming of Jesus will exhibit a slightly lower intensity of this at the end of the world.[14] This is the same presence of Jesus who did this at Creation on Days 1 and 2. By the end of Day 2, the atmosphere was fully developed as a result of the vaporization. It provided a shield from the intensity of the light being shone by Jesus, and later from the sun. The second coming of Jesus shows that His light can be directed to have different effects on different strata. The wicked will be destroyed by the brightness of His coming,[15] while the righteous will see Him and rejoice.[16]

Science shows that light is a natural by-product of heat. Tremendous heat is required to melt ice and the other elements on the earth and vaporize them. The Bible speaks of fire that will come down from God out of heaven and devour those who would not be saved.[17] This is the same fire or heat that caused creation ice to melt.

Some of the water and other elements turned to vapor, and stretched out around the rapidly forming sphere, producing what we call our atmosphere. The pressure of gravity pulling against the elements along with other forces heated the interior as the light from God heated the outside, and the earth formed with water covering the surface and filling the air.

The morning of Day 2 arrived with heat that finished melting the ice. The light reached every crevice of each melting filigree.

Gravity held the chunks of rock and molten ice together and drew them into a ball. This was probably the stormiest day of Creation, and quite likely the most tempestuous day the earth has ever seen.

Light and heat were not the only environmental issues involved. When a mass of cold air meets a mass of hot air, storms form. I was born and raised in Kansas. I'm familiar with these types of storms. On a sweltering hot day, if we heard that a cold front was approaching from Colorado, we knew there was a likelihood of tornadoes. If you're paying attention, they don't really sneak up and surprise you. You know when they are coming. On Day 2, there were probably tornadoes and other mighty storms all around the world.

There was no rain in these storms of Creation. "Neither wild plants nor grains were growing on the earth. For the Lord God had not yet sent rain to water the earth, and there were no people to cultivate the soil."[18] Rain didn't begin until the Flood more than a thousand years later.

SHAPING THE EARTH

Jesus was not the only person in the Godhead actively participating in Creation. "I (wisdom) was there." Some believe this is the Holy Spirit speaking. "I was there when He (Jesus) set the heavens in place, when He marked out the horizon on the face of the deep, established the clouds and fixed securely the fountains of the deep."[19] This horizon was developed by the melting elements, gravity, and hydraulic equilibrium. We will consider what the fountains of the deep are a little later.

Another step Jesus took to set the foundations was to give the sea its boundary. "Who kept the sea inside its boundaries as it burst from the womb, and as I clothed it with clouds and wrapped it in thick darkness? For I locked it behind barred gates, limiting its shores."[20] Job 38 is a great chapter in the Bible to give us insight into some of what had to be done at Creation. It also shows us some reasons God did these things.

Look at all that happened on Day 2 as God firmly established what the new shape of this earth would be. He set the limits. He set the foundations. He set the places where the water would be and where it would not be. He put the clouds in the sky. He established the atmosphere that we have. All of this was accomplished on Day 2.

Water played a key role in the first few days. It had to be strategically placed as part of the foundations of the earth because all life depends on it.

WHAT IS A FIRMAMENT?

We can describe the firmament as our atmosphere. Although completed during Day 2, water and element vaporization may have already begun forming on Day 1 when the heat caused the ice to begin to melt. As the mass of the elements in the air reached a certain level, enough was held in place by gravity so that the atmosphere formed. The atmosphere doesn't go floating off into space because gravity draws the molecules by their weight into the center and anchors them in place.

In order to leave the earth's atmosphere, a tremendous speed must be attained. Without the necessary thrust, it is impossible to escape the bounds of gravity. Rockets must achieve a certain speed before they can cast off the bonds of our atmosphere and break into outer space.

We know that hydrogen and helium gases are continuously being lost to our atmosphere. They travel faster than some other gases and are able little by little to escape.[21] Other elements were vaporized and became part of the atmosphere as Jesus moved from Day 1 to Day 2. "By the breath (or voice) of God ice is given, and the broad waters are frozen. Also with moisture He saturates the clouds."[22] This is a simple description of the development of the atmosphere.

OUR NEW GREENHOUSE

Today the "Greenhouse Effect" has a pejorative meaning in the discussion of climate change. Proponents blame the Greenhouse Effect for increasing temperatures on earth. But that was the intent all along. The gases in the atmosphere were held in by gravity and designed to keep heat in. That's the way God made it. Warmth began to normalize around the world due to the atmosphere developing to retain the heat and air.

THE GLORY OF GOD REVEALED

"But the Lord made the earth by his power, and he preserves it by his wisdom. With his own understanding he stretched out the heavens. When he speaks in the thunder, the heavens roar with rain. He causes the clouds to rise over the earth. He sends the lightning with the rain and releases the wind from his storehouses."[23]

God wonderfully designed the vapors that He caused to ascend from the ends of the earth to make our atmosphere. He made them on Day 2 because there would be things coming on Day 3 and later that would need the atmosphere to survive.

God knew in advance exactly what was required at every point in Creation week. He made sure everything was available before it was needed.

When God created, He did His work in an orderly fashion. He is a God of order. He also wants us to have orderly lives. The Apostle

Paul counsels us, "… be sure that everything is done properly and in order."[24]

In the same way, God made the world by His wisdom and discretion. He created the world in the appropriate order. There were reasons for the order of Creation.

Everything on Earth was put here at the appropriate time to serve something else that God would create later. The things that were first created were made for the benefit of that which would come afterward. Water would have many purposes. The first may have been to provide oxygen for the atmosphere. It also would be used by plants to grow, fish to swim, and animals and man to drink and be clean. The grass, trees, and herbs appeared in time to feed the cattle and other animals.

God's Creation order has a great significance that has slowly been lost sight of by many in our world. The fact that the earlier things came into view in preparation for the later works of Creation implies the greater significance of the living organisms produced later. Without this recognition, we can turn the Creation upside down. (See Chapter 7.)

For now, review Creation Week in reverse order, beginning with the creation of Adam and Eve. Before man, there were the animals, birds, and fish. We were given dominion over them. All of them were made for the benefit of man. Before the animals, birds, and fish were made, God created food for them to eat and land and water where they could live, firmament containing oxygen, so man and animals could breathe, and light to cheer and nourish their lives. He made the water, which could evaporate and fill the atmosphere with oxygen. He held the elements of the earth together by gravity to prevent them floating aimlessly through space.

The order of His Creation shows God's infinite wisdom. Natural laws indicate that God is a God of order, and since we are in His image, we may know He has given us orderly minds as well.

LOOK UP!

All the creation was special, but the heavens and the firmament have some exquisite glory. King David said, "The heavens declare the glory of God and the firmament shows His handiwork."[25] There's something remarkable about the skies that so affected the Psalmist that he singled out and praised those particular phenomena of Creation.

We might look up into the heavens and say, "Yep. There's air up there alright." But the Psalmist recognized through inspiration a uniqueness in the firmament and the heavens. While all of God's creation demonstrates His glory, the Psalmist chose the heavens above the rest. What is there in these aspects of Creation that we don't even comprehend? Why are these elements selected as examples of God's glory and handiwork?

If I were writing Psalm 19, my words might have been, "the genome and the reproductive system of man and woman declares the glory of God and the nervous system shows His handiwork." You can tell I studied biology for a couple of years before I decided God was leading me into ministry! When I examine Creation, those are the things that cause me to say, "Wow! That is amazing!"

Not only do the Psalms single out the heavens and the firmament, it also says, "the heavens declare His righteousness, and all the peoples see His glory."[26] When you consider the sky, you begin to see the immensity of God's righteousness.

The word "righteousness" means conformity to a particular set of expectations which vary from role to role. It describes the fulfillment of the expectations in any relationship. In other words, righteousness is your highest expectation of the relationship you have with someone. When there is nothing that could be better in any way, that is righteousness. Righteousness is accuracy. It is what is correct. It is the right thing. It is communal loyalty. It is salvation and well-being.[27]

The Psalmist writes that the heavens declare the righteousness of God. When you examine the heavens overhead and listen, God's Spirit speaks to you. Look at the sky from horizon to horizon and say to yourself, "This shows me the righteousness of God." When there is no way it could be improved, no way it could be better, that's righteousness. The heavens declare the righteousness of God.

The word righteousness in this sense also means that the heavens declare the salvation of God because He saves His people. The righteousness of God is seen in the heavens. By beholding we are drawn to God. By studying the heavens, we can be led by that righteousness to salvation. To all who are willing to see the creative work of Jesus Christ, the work of Creation opens the eyes of people to salvation for eternity.

That's why Paul tells us, "For ever since the world was created, people have seen the earth and sky. Through everything God made, they can clearly see his invisible qualities—his eternal power and divine nature. So they have no excuse for not knowing God."[28]

CREATION LEADS YOU TO GOD

Being led to such a study becomes an imperative. You can know God by seeing what He made, especially the heavens and the firmament. However, some people "knew God, but they wouldn't worship him as God or even give him thanks. And they began to think up foolish ideas of what God was like. As a result, their minds became dark and confused."[29]

It is clear that God speaks to His people in nature and in the things He created. We can recognize Him and learn of Him by studying His Creation, or we can choose to look away.

In 1992 I visited a primitive tribe near Ndoki in the Republic of Congo. This park has been described as the Last Eden and acclaimed for being the only place on earth where humans had not walked. Even the people who lived around it feared to trek into its depths.

As scientists researched what was there, they discovered the animal inhabitants had no fear of man, having never encountered them before.[30]

The remote Congolese tribe also had the stars. They had God's Spirit speaking to them saying, "Look at who God is and what He has done." God's Spirit could teach them through the starry heavens that God was Somebody who they could believe in and follow and serve. When viewing Creation, even the primitive can learn of God by the leading of the Holy Spirit. It is the same Spirit who leads others to a personal relationship with God through means of human communication or through God's written Word. No matter what our "textbook," the Holy Spirit will be sure the honest heart will not change "the truth of God for the lie, and worship and serve the creature rather than the Creator."[31]

Some have chosen to follow the God of Creation, while others have decided to make the things He created to be their gods. During all the ages of the earth, God has spoken to people through the heavens. Unfortunately, many cultures chose the heavens themselves to be their god rather than saying, "This shows me who God is."

They are without excuse because they have that which God has made, and they have the Spirit that speaks to them as well. Their origins are in the image of God. He reaches out to them just as He does to those in more "progressive" cultures.

This glory of God shown to us on Day 2 is a promise of salvation to all who worship Him as Creator. This is the meaning of "... worship Him who made heaven and earth, the sea and the springs of water."[32]

Remember, "When someone has been given much, much will be required in return."[33]

Some people live their entire lives excluded from access to verbal, written, printed, and digital declarations of God. They will not be expected by Him to grasp what may be considered the minutiae of details about God. They may never know His name, nor the theological fine points that "civilized" cultures fight wars over. Neither will they be penalized by God for what they could not have known.

The person reading this, however, whether by the tactile printed page, computer, or digital reader, has already learned enough to be summoned to further research to understand the complex simplicity of salvation by grace through faith in the sacrifice of Jesus, the Christ. Surrendering to Jesus will lead us to worship God as Creator.

God is calling on us to worship Him who made these things—not to worship these things. He wants to remind us that we need to worship Him in righteousness for salvation. We can learn of Him by studying what He has created. We will see in Chapter 8 that each of us was created in the image of God. This means we already have a connection to God and the inclination toward Him to see him, know Him, and long to worship Him. He is calling us back to our original foundations.

We are confined in this box of time with the beast who seeks to destroy us. His key instrument is to turn us away from our Creator. Our key defense is to stay close to our Creator God.

Chapter 3

Creation Day 2, Part B

GIVE ME ANOTHER GLASS OF THAT WATER

"Then God said, "Let there be a firmament in the midst of the waters, and let it divide the waters from the waters. Thus God made the
firmament, and divided the waters which were under the firmament
from the waters which were above the firmament; and it was so."[1]

By the end of Day 2, the ice had all melted. The gravitational
force locked the atmosphere in place. The water in the air and that
on the ground was separated by a visible difference between water
and gas. These waters didn't intermingle again until the flood of
Noah's day. After the Flood, we see evidence of another dividing of
the waters similar what occurred on Day 1 and Day 2. Creation is
key to understanding where it came from.

"When Noah was 600 years old… all the underground waters
erupted from the earth, and the rain fell in mighty torrents from the
sky. The rain continued to fall for forty days and forty nights."[2]

What kind of water do you see here? You see rain that fell for the
first time. But that's not all. The fountains of the great deep broke
loose and gushed out onto the surface of the earth. What are the
fountains of the deep?

A man purchases an undeveloped property. One of the first
things he does to improve this new piece of land is to set up a drill-

ing rig to find water. Depending on the location chosen for drilling, he will most likely find water at a certain number of feet down in the ground. The well for the property supplies water for both the family needs and for the animals to drink. Well water can be found just a short distance from the surface of the ground in many locations. It does not come from the deep.

The fountains of the great deep were made by God. Science has taken 6,000 years to figure it out. On the first day of the Flood, about 1,600 years after Day 2[3], everything reversed because of sin. The water in the sky once again met the earth in liquid form.

Until that day, rain never fell on the earth. At Creation, God divided the waters. At the time of the Flood, they met again. The waters were no longer divided from the waters. The windows of heaven were opened. The treasury of heaven lost its precious gem. Also, the fountains of the deep were broken up. There was water coming up from under the ground—from the "great deep."[4] This water was not described in Genesis 1. What was it, and where did it come from?

We discussed that on Day 1, as the ice melted rapidly, the water was pulled by the gravitational forces to find a place in the center of the earth. Some of the water moved to the interior of the earth and was trapped in these deep recesses. This is the water that gushed forth at the time of the flood.

Many have been perplexed attempting to explain the massive quantities of water necessary to cover the surface of the earth during the flood. I listened to one radio preacher trying to help God out with an explanation about the Flood. He described the source of the flood water by attributing it to evaporation of the waters of the oceans. "This," he declared, "caused the waters to take up more space than liquid water. Then it rose into the air and joined the other rain. That's how there was enough water to cover the earth."

Unfortunately for his explanation, once the evaporated water returned to its liquid state and fell back to the earth, it would only have replaced the water that had been taken from the surface through that evaporation. It didn't add more. Still, it was an attempt to support Scripture.

GOOD ANSWERS BEGIN WITH GOOD QUESTIONS

During my initial study of Day 2, I was perplexed like the radio preacher. I believe the Bible. I'm teaching it. I even believe the parts that have not been very popular in the scientific community and with those who scoff at Scripture.

An important Bible study principle is this: Don't just read a passage and think what everyone else says and then go on. Ask questions. Ask questions about every chapter, every verse, every thought, every idea, every word that you see in Scripture. Ask questions! Often the answer will come from a comprehensive understanding of the issue. It may not be enough to see only a phrase or verse. Being too narrow in our study often leads us to inaccurate conclusions. Then, be sure to listen to the Spirit of God speak to you with answers and more questions.

Let me give you an example: In the six hundredth year of Noah's life, all the fountains of the great deep, came to the surface. "All the fountains of the great deep were broken up and the windows of heaven were opened and the rain was on the earth 40 days and 40 nights."[5]

What was the result of all this water? "The waters rose and increased greatly on the earth, and the ark floated on the surface of the water. They rose greatly on the earth, and all the high mountains under the entire heavens were covered."[6] The waters rose higher and higher above the ground. The boat floated safely on the surface.

That's a lot of water. It had to be here from Creation either in the atmosphere or on the surface or somewhere. Where did it come from and where did it go?

Seeking an answer, on October 11, 2013, I emailed Rich, a good friend who is, among other things, a scientist. I referenced Genesis 1:2-4. I wrote:

"We see that in the beginning that there was water on the earth, as found in the term the face of the deep and 'over the face of the waters.'

"On the second day of Creation the waters, the upper and the lower waters, were divided by the firmament in between. This water of verse 2 is the deep, atmospheric water, some of which became water on the ground. So why didn't it all become water on the ground?

"Is it possible that all this atmospheric water in verse 2 was in some form other than liquid or vapor or solid, (such as ice)? Are we certain it couldn't be some other type of solid? Could it? Could it have been water, but not in a form we are familiar with? Could it have been combined with some other element that kept it from being what we call water? Might it be possible that the light on the second day caused some kind of reaction that either took away one or more of the molecules or elements of making liquid water, or allowing the combination of hydrogen and the two oxygen molecules to make water?"

A lot of questions went through my mind as I wrestled with where this water was. Rich had mentioned to me that the H_2O molecules that make water may have been created in the process and separated in the firmament.

My questions continued.

"Would this be saying that the water that was above and below the firmament may not have been in the form of water, or H_2O? It may have been pre-water. What is pre-water? Would it be water with the hydrogen and the oxygens not yet defined? Could this ball of rock be surrounded by or maybe somehow infused by water in some form? We might think of it as being somehow different from liquid or vapor. Perhaps it was pre-water with hydrogen and oxygen floating around but not yet brought together."

My mind works in strange ways at times!

Rich's answers to me were basically no, maybe, no, and we don't know! Of course, his answers were a lot longer than that, but I've summarized them. Rich said there really were no answers at that time.

SCIENCE IS LEARNING (SOME OF) THE ANSWERS

It wasn't until June 2014 that the answer was made known. Did you hear about it?

Northwestern University geophysicist Steve Jacobson and the University of New Mexico seismologist Brandon Schmandt found deep pockets of magma located beneath the continent of North America.[7]

Their discovery came by connecting 2,000 seismometers by computer and analyzing the information.[8] They couldn't have done this just a few years previously. The necessary equipment didn't exist then. Science is beginning to catch up with the Bible. A good many recent discoveries reinforce the Bible accounts of Creation and the Flood.

Jacobson and Schmandt began to hear what seemed like seismic ocean waves deep beneath the surface of the earth.[9] Ocean waves! How can that be? They connected their seismographs and analyzed the data. They found sounds of underground ocean waves across the North American continent.[10]

The report tells us, "This water is not in a form familiar to us. It is not liquid, solid, or vapor. The weight of 250 miles of solid rock creates such high pressure along with temperatures above 2,000 degrees Fahrenheit, that a water molecule splits to form a hydroxyl radical (OH), Oxygen Hydrogen which can be bound into a minerals crystal structure."[11] The scientific discovery jarred my memory of the questions in my emails to Rich.

How much water is under the earth's surface? Scientists estimate that the amount of water 250-400 miles under the surface is three times as much as all the water that is on the surface.[12] This is enough to cover the mountains.

This doesn't mean the giant Himalayan mountain range and other high peaks were covered at their present height. Most likely, these mountains grew a great deal under the conditions of the Flood. "The discovery suggests water from the Earth's surface can be driven to such great depths by plate tectonics, eventually causing partial melting of the rocks found deep in the mantle."[13]

It is easy to see how such dramatic movement of oceans of water in a very short time could cause substantial upheaval of the tectonic plates. Water moving out of the interior rapidly, and then back into its hideaway within a short time may have played a major role in pushing our high peaks up to near their present positions. This could also be another explanation for "ocean marine fossils at the top of these mountains" as "one of the key pieces of evidence cited that advanced the idea of plate tectonics."[14] Of course, NASA doesn't associate the fossils with the flood.

Even at that, Dr. Jacobsen mentions, "We should be grateful for this deep reservoir. If it wasn't there, it would be on the surface of the Earth, and mountain tops would be the only land poking out."[15]

Previously, science believed that the water on Earth was brought here by comets that struck the planet millions of years ago (such as the 67P Comet. See Chapter 2). This new discovery is "good evidence the Earth's water came from within."[16] They don't seem to have considered our theory stated earlier. Water was created here and may have traveled to other places on fragments that escaped the pull of our gravity when all the foundational elements of Earth were cast into space.

I applaud science that they are proving that it was possible for the Flood to cover the entire earth, just as the Bible says. It told us thousands of years ago that the water would gush up from the deep. Later, when it was no longer necessary, the extra water found its way back down into its place in the deep.

PATIENCE AND PERSEVERANCE

When we look at Day 2, we can see that God exhibited patience and perseverance to make the framework of a world that would sustain us. He knows the future because He is in the future. We talked about that on Day 1. He knew exactly what He would need and where He would need it.

He built those same characteristics in people. He built into us patience and perseverance that lead us to take time to construct the framework for what needs to be done. If we choose, we can look to the future and say, "I'm going to need this someday, so I'm going to build it well now."

Because He built those characteristics in you, find these things in yourself. Find them. They are there because He made you in His image. Those things that He is like, He created in you. Practice them and allow God to renew them in you to become more fully in His image.

WORTHY OF WORSHIP

God's created world, including the way He made us, draws us to worship Him. "I will praise You for I am fearfully and wonderfully made. Marvelous are your works and that my soul knows very well."[17] None of the man-made gods of the earth have done what this God has done. He is the living God to whom we say, "Who would not fear you, O King of nations? That title belongs to you alone! Among all the wise people of the earth and in all the kingdoms of the world, there is no one like you."[18] He is the only one who "made the earth by his power, and he preserves it by his wisdom. With his own understanding he stretched out the heavens."[19]

Many things are called gods. There are gods of wood and stone, precious metals, activities, and ideas. They cannot see, they cannot hear, they cannot help those who worship them. Indeed, they are made by people. They have never created anything, nor can they.[20]

SCIENCE CATCHES UP

Nearly four thousand years ago, the Bible told us about the fountains of the deep. Now science is finally catching up! We need not be concerned about scoffers who discredit the Bible based on science, politics, culture, or any other philosophy or field of study. All these platforms are real, and all have their place in the discussion. Yet, they all have their foundations in assumptions. Far too often these assumptions were established specifically to discredit God, the Bible, and/or their followers.

This was even foretold. "I want to remind you that in the last days scoffers will come, mocking the truth and following their own desires. They will say, 'What happened to the promise that Jesus is coming again? From before the times of our ancestors, everything has remained the same since the world was first created.' They deliberately forget that God made the heavens long ago by the word of his command, and he brought the earth out from the water and surrounded it with water."[21]

Don't allow scoffers to turn you from who God is and what He tells us. Neither can we make the mistake of ignoring true science. This has been done in the past to the discredit of the church.

THE NEED FOR BALANCE

The water cycle and the dividing of waters on Day 2 involves more than just the water that circulates between the atmosphere, the oceans, and other surface waters on the earth. It extends deep into the interior of the planet.

You know what happens in earthquakes. The crust of the earth moves. Some have asked, "How does that affect the delicate balance of the earth to keep it from spinning out of control and wobbling right out of its orbit?" There may need to be certain amounts of wobble to maintain equilibrium, but the earth is constantly changing.

Put this in terms we can experience for ourselves. When something goes wrong with a tire on your car, you can start having a bumpy ride. The mechanic tells you that your tire is out of balance. For whatever reason, something caused one side of the tire to be a little heavier than the other side. To resolve this, the technician adds another weight in the appropriate place to prevent the bumpy ride or even an accident.

That's what happens in earthquakes. The tectonic plates move from one location to another as the earth is rotating. Everything needs to be balanced for the earth to have a smooth rotation and orbit. Perhaps someday we will learn that when the tectonic plates move, the water 250-400 miles below the surface moves into the spaces left open by the movement of the tectonic plates and rebalances the earth. God may have planned for this to happen. The same would apply if an imbalance of surface water occurred due to cyclical raising or lowering sea levels and diminishing or increasing arctic ice. Here is more insight into the character of God.

God always maintains proper balance, and He created our earth to keep this proper balance as well. We, who are made in His image, were also formed to have well-balanced lives. There must be balance in our physical life. For proper health, we need to have balance or moderation between the things we eat and don't eat, exercise, rest and work. Just as our physical lives need balance, there must be balance in our spiritual lives.

Since the Fall of Man, there has always been a problem keeping a proper balance in the spiritual life. There is that straight way that God has marked out. We are warned not to go to the right or the left. But the devil calls from either side saying, "Come this way—or go that way—I don't care which way you go. Just don't stay straight."

People get out of balance. There are many nomenclatures for imbalance. What you call it doesn't matter. As soon as we get off the path that God has set for us, we become unbalanced in our lives and cannot accomplish what God wants us to achieve. The balance in our spiritual lives comes from having the Water of Life hidden deep within our hearts.

ARE YOU THIRSTY?

What is that Water of Life? Or should we ask, "Who is the Water of Life?" Jesus gave the answer. "If anyone thirsts, let him come to Me and drink."[22] To the woman at Jacob's well, in Samaria, He said, "But those who drink the water I give will never be thirsty again. It becomes a fresh, bubbling spring within them, giving them eternal life."[23]

Even as Jesus created this earth, He became the provision for us to have the Water of Life that He knew we would need. He created water because without water, He knew we would die physically. Without the Water of Life, we will die spiritually and eternally. Even though Jesus created the water, and even though He, Himself, was the Water of Life, when He hung on the cross, He cried out, "I am thirsty!"[24] He created water knowing in advance He would die thirsty. Through His sacrifice, He liberally provided us with the Water of Life that we need never thirst again.

INVISIBLE DOES NOT EQUAL UNIMPORTANT!

Our firmament tells us that even the things we can't see are essential to the existence of the whole. Have you ever seen God? The fact that God is invisible causes some people to doubt His existence. But we can't see the air—yet we know that we depend on it for our very lives. God can't be seen, but we rely on Him for our physical, mental, emotional, and spiritual lives.

Some believe that their true inner selves are invisible to those around them. Frankly, others probably know more than we might like for them to know! But who are you, really? Can anyone know all about you just by watching you walk down the street or sit on a park bench? The terrorist who is about to blow himself up in a crowded market often can hide who he really is, and what is inside the mask he wears. But soon the secret he covers up will destroy him, and perhaps others around him.

When Jesus walked the earth, did anyone know who He really was? The Bible says of Jesus that there was nothing beautiful or majestic about his appearance, nothing to attract us to him.[25] Who

50 • GARY WAGNER

Jesus was on the inside could not be easily seen from looking at the outside.

God created each of us in His image to be good, holy, honest, creative, and to have many other positive characteristics. Each one must answer, "Is that who I am now? Really?" Am I the clothes I wear or the way I comb my hair? Am I the friends I keep? Do the things I do when someone I like is watching me show who I am?" Probably not.

When you are alone, you might be a person who reads your Bible daily and totally surrenders your life to God. Or you might be one who goes into your room and shuts the door and watches Internet porn for hours instead of sleeping or studying. You might struggle with your sexual identity. You might be angry at someone else. You might be depressed or have an eating disorder. You might be a cutter, or an alcoholic, or a druggie. Are you happy with who you are?

No matter what you appear to be on the outside, what you really are is who you are on the inside—the part other people can't see. You may choose that part of you to be something other than the image of your Maker. If that is your choice, however, you will hurt yourself and those around you who love you. You can't change the inside. You can only choose to allow God to change you.

God created you to be perfect. But there is an enemy who hates goodness in God and in you. He is the beast who is captured in this box of time with you. The evil one is responsible for all the suffering in this world. He had planned before you were born just how to destroy you. He works earnestly to lead people to take just one more step away from the image of God within them. As a result, individuals and societies do the most terrible things. Perhaps you have too.

In many cases, people don't even realize how awful the things they do are. They are just following the example of those who lived before them. But to live a life apart from God will ultimately destroy all who choose it. That's what the evil one wants. Whether we realize what we do is wrong or not, it will still hurt us. If you take cyanide, you will die, whether or not you recognize cyanide is what you are taking.

God wants to recreate us so that we will again be in His image of goodness and righteousness. He created you perfect in the beginning. He can recreate you perfect again. You can't do it for yourself. You can only choose to let Him do it for you.

GOD'S WORK IS PERFECT

We need to understand God has not given up on us and He never will. If we had seen the world at the beginning of Day 2, we would have thought there was no hope of it becoming anything worthwhile. But God didn't stop creating until it was finished. It is the same for you and for me. The same Creator God who made you will not stop recreating you until you are finished.[26]

On Day 2, the earth was perfect for what God intended it to be at that time. Even though it was not possible for the earth to sustain the life we are familiar with, it was rapidly becoming capable of supporting life. The moment any person surrenders their will to God, he or she is declared by God to be perfect. That person is then ready to allow God to finish the work of rebuilding a truly perfect character in them. This is what renews the image of God in them. In the same way, the earth was imperfect (or at least incomplete) on Day 2. Yet it was exactly what it needed to be for that time in its existence.

If you have surrendered yourself to God, then you, too, are exactly the way you are supposed to be at this point in your life. The recreation process will be up to God if you allow Him to make the other changes needed. These changes prepare you for the final scene when Jesus comes to collect His people and take them home. You, too, are perfect today if you allow God to do in you His perfect work of returning you to His own image. And tomorrow, He will make you more perfect.

The earth had nothing to do with its own creation, and you and I can't change ourselves. You cannot quit doing the things that you know you should not be doing. It's not your job to stop doing them. It is God's job to remove them from you if you are willing to let Him do it. We must allow God to make the changes that need done in us.

"For God is working in you, giving you the desire and the power to do what pleases him."[27] He gives you the desire for His goodness to be worked out in you, and He offers you the ability then for that to be done because it is He who does it.

Do you feel you will never be the person God wants you to be? Take hope. Remember the earth without form and void at the beginning of Day 1. Then claim the promises of God.

"And I am certain that God, who began the good work within you, will continue his work until it is finally finished on the day when Christ Jesus returns."[28]

As we come to the end of Day 2, the evening and the morning of the second day, do you hear God say, "It is good." Don't continue to harass yourself that you have areas of life that need changed. Surrender to Jesus. Allow Him to take away those things that plague you. Allow Him to finish the work of Creation He longs to do in your life.

This is the lesson we must learn to be safe from the attacks of the beast in the timebox.

Chapter 4

Creation, Day 3

WHAT A DIFFERENCE A DAY MAKES

At the beginning of Day 3, imagine the earth as a quagmire of seething sludge, mixed together in a spherical cauldron of boiling mud. Evolutionists talk about primordial soup. But this "soup" wasn't exactly the way the evolutionists describe it.

Since the Bible specifies that each evening and morning comprise a day, the evening of Day 3, like Day 1 and Day 2, was the first part of the 24-hour cycle. From Creation until today, each biblical day began at sunset. Even in our society, we count a new day as beginning at midnight during the dark part of the day.

When the third day began, the atmosphere formed on Day 2 surrounded the earth and held in the heat that came from the energy source, which was Jesus. The air temperature stayed warm, instead of the heat escaping. The atmosphere caused the greenhouse effect, as noted in Day 2. The air must have been steaming by evening because of the tremendous heat required to melt the ice during Day 2. Cooling began that evening, at the beginning of Day 3.

As the Earth completed its rotation on its axis, morning drew on and the Light was there right on time. The Light of the World revealed a globe of mud. The sky, a field of beautiful blue, rested above all the earth to show the promise of the glory and wonder. The world was not yet conducive to any of the amazing life forms that God planned to put on it.

ORDER OUT OF CHAOS

Ice had melted on two levels. Some ice became water and flowed in great flooding rivers to the center of the gravitational pull. As this swirling mass moved toward the center of the planet, it carried with it stones and minerals, the elements, the dirt, the precious gems, the gold, and all the other earthy ingredients. Deposits were made along the way.

Here is the Bible description of Day 3. "Then God said, 'Let the waters beneath the sky flow together into one place, so dry ground may appear.' And that is what happened. God called the dry ground 'land' and the waters 'seas.' And God saw that it was good."[1]

Day 3 saw the completion of the construction of the foundation of our planet. The infrastructure for the establishment of new life had been strategized and completed. Life was built on solid rock. The rock would last forever, except by the direct intervention of God. The rock and the water contained all the nutrition and hydration that would be needed to keep the coming flora and fauna alive throughout eternity. It didn't look like much, but this newly-formed fortress was a treasure house of all the wealth that would be sought and fought for by kings and kingdoms throughout its future.

The implication is simple. We are to build our lives on the Rock—the foundation, which is Jesus. "For no one can lay any foundation other than the one we already have—Jesus Christ."[2] He is the Rock, and He is the Water of Life. "Those who drink the water I give will never be thirsty again. It becomes a fresh, bubbling spring within them, giving them eternal life."[3]

As we move into the dramatic change of Day 3, keep an eye on the repetition of words and phrases used to describe the process of creating life. They will be our clues to dig for greater meaning that is hidden in them.

Imagine viewing the glamour and drama of the next incredible step toward beauty and completeness. "Then God said, 'Let the land produce vegetation: seed-bearing plants and trees on the land that bear fruit with seed in it, according to their various kinds.' And it was so. The land produced vegetation: plants bearing seed according to their kinds and trees bearing fruit with seed in it according to their kinds. And God saw that it was good. And there was evening, and there was morning—the third day."[4]

LET'S PLAY IN THE MUD

This must have been a fun day for Jesus. Days 1 and 2 were all dirty construction work. But on Day 3, I can picture Him with the exuberance of a small boy and the finesse of the most talented artist joyfully beginning to do His day's handiwork. He knew exactly what He was going to create. The Bible says. "For when he spoke, the world began! It appeared at his command."[5] I agree that's the case. Still, I can't help but think that Jesus also had some fun with Creation. Throughout Genesis 1, when He spoke, "This is what I'm going to make…" ("Let there be…" the Bible says). It adds, "And then He made it."

Picture Him bending over this vast sphere of seething mud, as it cooled down. Watch, as with a sweep of His arm, Jesus forced the mountains out of the rocks and dirt. Can you picture a little boy intensely peering into that great orb of ick? He lifts his arm and dips it in and pulls the dirt up, and makes mounds of dirt in some places, leaving other deep areas where the water gathered.

I like to think of Jesus doing that! All around Him, the earth, the land, and the seas took form. "Let the waters under the heavens be gathered together in one place and let the dry land appear," the Bible said. "And it was so."[6] But the day wasn't finished! Beauty was about to overcome the drab.

Next, when the dry land was gathered here, and the waters were over there, Jesus said, "Let the earth bring forth grass."[7] Watch as Jesus waves His arm over that muddy mess. On one side of His extended arm, it's still mud while on the other side, grass sprouts up instantly. Can you see the green blades pushing through the brown, oozing dirt? As His arm crossed the panorama, grass shows up. And not just the grass, but flowers, trees, and everything beautiful. All the vegetation that had never existed. Suddenly it appeared. It just showed up. What a difference one day can make! A molten mass of mud transformed in a single day to water and land, beautiful green fields and hillsides, and blue skies.

"And out of the ground the Lord God made every tree grow that is pleasant to the sight and good for food. The tree of life was also in the midst of the garden, and the tree of the knowledge of good and evil."[8] This describes some of the things that He put in place.

Now, Jesus didn't just make all this up as He went. He planned it out carefully. It didn't happen accidentally. He didn't say, "Let it be," and whatever wanted to happen, happened. The earth couldn't have made itself in that way. No, Jesus knew exactly where every blade of grass would grow, in the same way as He knows every hair on your head. And He remembers the ones that you've lost!

Jesus knew just what shape every mango would have, how every pineapple would taste. He pictured you and me as we sink our teeth into the juicy flesh and savor each succulent bite. And then Jesus lifts

His finger to wipe away the juice that dribbles down our chins. With a flourish, He hands us another slice and says, "Enjoy some more."

If we had been present at Creation, we might have missed seeing all the miracles. So much happening, over such a vast expanse of earth. Stimulated by everything taking place around us, our eyes and our minds would dart from one thing to the next. "Wow!" Did we miss the fact that every little blade of grass, every tree, and every flower was a miracle made by Jesus? Do you overlook those same miracles in the frantic pace of your life today?

What a difference a day makes. God designed everything with you and me in mind. He produced it perfect just for us. Do you enjoy looking at it? He made that for you. Anything you like to see, He formed that knowing that you would find pleasure in it.

THE WONDER OF COLOR

God could have created our world in gray and white! But instead, He created color. It is more than just to be nice to look at. Different colors relate respectively to the body, the mind, the emotions and the essential balance between these three.[9] Every color has an effect on you. Science reports the most relaxing colors are the cool shades of blue and green—the colors you see most in the sky and on the ground.

Researchers in neurology, psychology, and ophthalmology have found preliminary evidence that the relaxing effect of green does not depend on cultural associations with leaves or meadows.[10] In other words, the reason you like green and find it soothing isn't just because that's what you grew up with, so that's what you like to see. No, there are bigger reasons than that. God knew what you needed because He knew the way He would create you.

"Our eyes perceive colors using tiny sensors called cones. Certain cones are sensitive to red, green, or blue light. However, overall most are sensitive to wavelengths at 510 nm, which translates into green light. Researchers hypothesize that this sensitivity to green objects might affect hormonal production or the circulation of neurotransmitters that in turn influence moods."[11]

Using different colors has even been used to "heal medical problems. Chromotherapy goes back to ancient Chinese and Egyptian practices. Recent studies have shown that when exposed to green-colored paper or placed in a green room, a person's heart rate drops, the blood pressure lowers, and muscles relax. Hot colors, like red and orange, cause blood pressure to rise."[12] What are the colors you have in your bedroom, the great room, or your family room at home? Many people have different colors. Sometimes it speaks to the way they feel about life. Perhaps it causes them to feel the way they feel about life.[13]

You see, God knew exactly what He was doing when He made the colors of this earth. Perhaps He used other colors in other locations in the universe—even colors that we have never seen or imagined. Maybe the spectrum is different there. But in our world, Jesus knew exactly how our eyes would function and how all color would affect us.

GOD SAID IT FIRST WITH FLOWERS

Just think of this: "God was not content to provide what would suffice for mere existence. He has filled the earth and the sky with glimpses of beauty to tell you of His love for you."[14] God was the One Who first thought of giving flowers to say, "I love you!"

All those flowers, all the birds that fly—the beautiful way that they fly and their varied colors—were placed here by God to remind you of His love for you.

During a trip to Australia, I observed many bird species I had never seen before. The vibrant colors of their plumage dazzled my senses. What a fantastic place to see birds that you can't even imagine. God created magnificent creatures to demonstrate His love for us. You could have lived your entire life without knowing that a bird or a flower could exist. But God placed them on Planet Earth because He loves you and wants you to enjoy them.

GOD'S PANTRY

There's something else important about these plants. "Then God said, "Let the land sprout with vegetation—every sort of seed-bearing plant, and trees that grow seed-bearing fruit. These seeds will then produce the kinds of plants and trees from which they came.""[15] And that is what happened.

Other than looking good, what was the reason for this vegetation? Do you think God put it there only to cover the ground to prevent erosion when it rained? No. Among other things, He said it would be for food. Food for whom?

God said, "Look! I have given you every seed-bearing plant throughout the earth and all the fruit trees for your food. And I have given every green plant as food for all the wild animals, the birds in the sky, and the small animals that scurry along the ground—everything that has life."[16]

This was the first diet! Do you notice anything not included in that list? There are no vegetables! Do you like vegetables? When Mom says to the kids, "You must eat your vegetables," they might reply, "Adam and Eve didn't have to eat vegetables!" (That's just for those of you who don't like vegetables.)

God knew on Day 3 that you and I would need something to eat, and He made it for us before we were created, so it would be here when needed. When the first lunch was called, there was something ready to eat.

SEEDS ARE A P-R-O-M-I-S-E

This is another case of common words or phrases being used multiple times in a short space in Scripture. It tells us, "This is important!" In the description of Day 3, we see the word "seed" used repeatedly!

God made things in ways that were specifically needed for us long before we knew that we needed them. Why were there seeds in everything that He mentioned? To grow new plants and trees.

If sin didn't exist, there would have been no death. Without sin and death, new plants would not need to grow. God knew that man would sin. He knew sin meant death. Death upended God's intention of His creation and creatures—man, animals, and plants—living forever. Knowing what was going to happen, God implemented a course of action that dealt with sin and death before they even existed.

God's plan would take at least 6,000[17] years for it to come to fruition. Which brings us about to our time. Perhaps His plan will take longer. We don't know. But God knew the earth would become a dismal place once death took over. He placed within all living things a method for regeneration.

Do angels have babies? No. Why? They have never sinned, so they never die. There is no need for them to regenerate. Everything on our earth has seeds to be able to grow new things. Do you see the glory of everything that God made here on the earth? He knew what was going to happen, and He made provision for the continuation of life.

On Day 3, God placed seeds in plants and scattered them throughout the entire earth. He wanted to ensure plants continued to exist, even after sin and death invaded the planet. Seeds that we take for granted are a promise to us that life will renew itself until God's plan to restore us is fulfilled.

We may overlook seeds. We look at them and think, "That's just a seed." No. That's a promise from God! "Don't worry," the seed says. "God has not left us without our daily provision." The unpretentious seeds provide more than just physical food. Seeds remind us to "seek first his kingdom and his righteousness, and all these things will be given to you as well."[18] Seeds are a promise of God's provision for eternal life spiritually as well as physically.

CONSIDER THE TREES

God also created trees on Day 3. Have you ever pondered a tree? Most people worldwide are used to trees because we see them daily.

When my family first moved to New England after living in Kansas, we were not used to so many trees dotting the landscape. In Kansas, we were used to seeing for miles across the plains. We moved to New Hampshire in late summer when the trees were at the height of their summer foliage.

Trees stood as sentinels, lining both sides of nearly every road we drove! With the arrival of fall, the once-green leaves turned to yellow, orange, red, and burnt umbers. Soon they released their hold on the branches and fell to the ground. As the trees stood naked against the fall chill, we discovered during the past two months, we'd regularly driven right by a mall and Walmart without even knowing they were there.

What would happen if you were seeing trees for the very first time? They are stunning feats of engineering. How can they stand so tall and stretch out so much weight? That's hard to duplicate. Structural engineers today can't do it. But all of this was for more than shade and beauty.

Revelation 22:2 tells us about trees in Heaven. "In the middle of its street, and on either side of the river, was the tree of life…"[19] The Greek word for tree here is "ksoolon."[20] It is the same Greek word in Galatians 3:13, "Christ has redeemed us from the curse of the law, having become a curse for us (for it is written, 'Cursed is everyone who hangs on a tree')."[21] It is talking about the cross on which Jesus died.

Jesus made His lovely trees to be a tool man would use to kill Him. Can you see the irony of that? He made the seed so that everything would continue to grow. On the same day, He made the instrument that would be used to give us victory over sin. On Day 3 we see grace!

God had arranged for our salvation before we became sinners. If the cross is an instrument of grace, then so is the Tree of Life. Even the Tree of the Knowledge of Good and Evil is a tool of grace! It is the symbol of your freedom and mine to make our own choices. Its presence in the Garden is God speaking to you. He is saying, "I will let you decide whether to serve Me or not."

The beast with whom we share the box of time was given this tree as his point of access. He had chosen to become a beast because, as with all of God's creation, he had the power of choice. And confined to his tree, he was eventually able to make contact with Adam and Eve, who chose to listen to his lies.

You also have the freedom to choose. God knew exactly what He was doing and how it would affect everything that would come afterward. By dying on the cross, He brought beauty and glory out of what seemed like chaos and the turmoil of sin. This is exactly what He did on Day 3. It was evident by His exquisite planning that He could see the future, and whenever His way was followed, everything would be good.

Earth was made to last forever. He placed the self-perpetuating seeds of growth and eternal beauty. God intended man to last forever, too. God made you to last forever because He wanted you, and He thought of you before Creation. He thought of you, and He said, I want this one to be with me forever. He put within you everything that was necessary for you to be able to live eternally.

But, sin came into the world and changed everything. It brought suffering and death into your life and mine. We know that we will die. God's plan for us is that even though death is in our future, we have a place with Him in heaven. God designed this excellent strategy, which will continue until it is completed. You and I have the power to choose if we will follow God's plan, or institute our own plans that can only lead to eternal death. "There is a way that appears to be right, but in the end it leads to death."[22]

POWER TO CREATE

You remember that we were created in the image of God. Let's learn from God's work of Creation to understand better what He has in store for us.

God gave you the power to create. How do we create? One way is with our words. You create with your words daily. Already in your life, you have created with your words. God observes to see what you are building. Ask yourself, "What am I constructing with my words?"

Who are the people you know who can make you do anything with their words? In their presence, you are relaxed. You feel better about yourself and about your worth. You do things that you would never dream of doing, other than the fact that they asked you to do them.

On the other hand, think of the bully. What does he create with his words? Bullies make ruined lives. Think of the alcoholic mother of small children, who, if she disciplines at all, only yells at her children. Those children are likely to grow up to be the same as their mother. She has created them.

What do you create with your spoken words? Think about it for a moment. What about gossip? That is certainly a way we can create something out of nothing! Aren't we accountable if we create broken spirits and lives with our gossip and backbiting?

If you aren't a Christian, you can skip this paragraph. What about the one who neglects to tell others about Jesus? Are you using your creative words in the right way if you refuse God's command to tell others about Him? Are you the person who shares the gospel of Jesus with others in a positive way? Do you point others to a loving, saving God? Are you the one who shares a kind word, who uplifts people in their troubled times? You are creating with your words. You can create sorrow, pain, and distress. Or you can create joy, hope, and peace. God gave you that power.

Remember the parable Jesus told about the talents. He said the kingdom of heaven is like a man who took a trip to a far country. Before the man left home, he distributed different amounts of silver, among his servants.

When the man came back, he asked, "What have you done with what I have given you?" The one who had five bags of silver had done much. "Well done, my faithful servant."

The servant who had only one bag may have felt, "I'm not as good as these others, so the same thing certainly will not be expected of me. Besides, I know the master is harsh, so I will just give back to him what he gave to me." But the master said, "Take from him what I have given him and throw him into outer darkness where there will be weeping and gnashing of teeth."[23]

What if you have only one talent? How are you using it? What are you creating with the talents God has given to you?

ONE SINGLE DAY

What difference does a day make in your life? That depends on you. You see the difference a day made on Day 3 when God took Creation from a blob of molten mud to a place of beauty and great glory. What are you doing with a day in your life? Any day? Every day. What are you doing with the life you are now living, whether good or bad (or a mix of good and bad)? The life that you live now can be amazing. It is up to you. It's your choice. You are the one who determines what a difference each day can make for your own life and for the lives of other people.

SOLID FOUNDATIONS

Before God could make the earth beautiful, He had to lay the foundations. Foundations are crucial. Remember Day 1 and Day 2? On Day 1 God just put everything in space where it needed to be. He showed up with the light and the heat, and He began to melt it down. Day 2 completed the process and the atmosphere formed. Those things had to be there before He could put the grass on the ground, before He could plant the flowers on the rolling hills, and all the rest that was needed.

In your life, do you want to do something great for God? Do you want to be something significant to bring glory to God? Then build the foundations. If you're a student, this is your time to build the foundations. No matter what your age or situation, God wants to do beautiful, wonderful, and incredible things with your life. But He builds the foundations first. You can't construct magnificent structures with a crumbly foundation. They'll fall, just as the house of the foolish man who built his house on the sand.[25] You don't want to build on sand. Instead, ask God for wisdom to follow the example of the wise man and build a firm foundation on a rock. God is trying to create strong foundations in you. Let Him.

In our lives, the foundation is what we call character. God wants to build your character to be alluring and glorious and pointing back to Him. Then everyone who sees that appealing foundation of character will be looking at God and not at you.

If your character is still being built, some of what you are today may not seem beautiful. But as long as it's in the process of being completed, it's good. It is the same as early Creation. It was good at the end of every day. Not by comparison to the next day, but by what God expected it to be that day. By allowing God to continue to do His work in you, you will see the beauty and the eternity.

God knew what Adam and Eve would do in Eden's paradise. Just imagine. They had everything perfect. We sometimes complain because of the way things have played out in our lives. We look at them and say, "Oh, if only this," or "if only that." But reality is, even if we had everything perfect, we would probably choose the same actions as Adam and Eve. We would sin. God loved them and provided for their salvation even before they were created. He knew that they would sin. He did the same for you and for me.

Ask yourself, "If the seed is so important to teach us about God's provision for perpetual continuity of all living things, what is the effect of hybridization of seeds?" They grow only one generation and then must be replaced by more hybrid seeds. A big lesson here is that man is responsible for continued life. We can give the benefit of the doubt to science. Perhaps they are not aware that they are taking over the role of God to provide "what you will eat and what you will drink." This issue is expanded on in Chapter 6.

Every time you see a seed, praise God for providing for you continually until He comes. When you see a tree, let it remind you of the grace of God to save you and of your need day-by-day, moment-by-moment, to choose to be saved. It's all up to you. Creation points you to God's wonderful plan to make something captivating and perfect, not just for you to live in, but of your life. Something that will last throughout eternity. Something that you will be able to enjoy. Something that will enable you to glorify God. Something that will allow Him to be glorified through you. But you must make the choice.

Are you allowing God to build you in His image, starting with the foundations? Are you allowing Him to use you, to speak to you, and through you? Are you becoming the new creation He wants you to be? Is your character growing ready for eternity? I pray that you will accept God's re-creative work in your life. Will you be the person God created you to be?

Chapter 5

Creation, Day 4

NO DARK VALLEYS

In Phnom Penh, Cambodia, 1973, the Khmer Rouge[1] waged war in the countryside. As a 22-year-old college student, I came to Cambodia's capital city as a volunteer to lead out as director of an English language school and oversee relief efforts for the country's citizens affected by the war. Part of my job was to distribute relief supplies, clothes, and food to the refugees who had fled from their homes with nothing but the clothes on their backs.

EVACUATE NOW!

From time to time the United States Embassy would contact me with the message, "You need to get your people out of the country quickly. The Khmer Rouge are getting ready to attack Phnom Penh."

We were expected to close the school and send our American teachers out of the country. These evacuations happened four times while I was there, although I left Cambodia only once. We knew communist spies infiltrated the school as students to make lists of all attendees. Those lists would become execution lists if the Khmer Rouge took over. These communist sympathizers didn't like the fact that people would study a foreign language, especially English. They despised those who came to Bible classes and meetings. They especially hated those who became Christians and worshipped a foreign God. In their opinion, anyone who wore glasses, had dental work, or had a relationship with someone from a foreign country deserved to be on the execution list.

When required to evacuate from the country, I searched out every piece of paper in the school. I hand-carried the vital records with me to Saigon where we were safe. The mundane, less critical documents still had to be burned one piece of paper at a time. Execution lists included any name found on a school document.

A month later, with the immediate threat of invasion past, we returned to the school and started all over again. We resumed our work, knowing evacuation orders could come again any day.

After the first required evacuation, I decided to remain at the school by myself when the other staff members went to Saigon and Bangkok. Unless absolutely necessary, completely closing and re-opening the school required too much work. I decided to remain in Phnom Penh and take my chances.

AN ATTACK BY THE BEAST

One day, while working in my office on the ground floor of the building during an early afternoon rest time, I heard explosions. It was common to hear detonations because the country was at war. Immediately, I ran up to the roof, which was the fourth floor of the school. I gazed out over the city. Then I saw a couple of explosions in the southeast part of the town, then two more to the west, with continuing sets westward. The Cambodian military used American 105 mm howitzers to defend the city. The Army fired on the Khmer Rouge. The Khmer Rouge counter-attacked the military.

The relentless attacks of the rebels forced the poorly trained Cambodian troops to desert their posts. They left behind their artillery. They didn't disable the howitzers, they just fled. The Khmer Rouge turned the artillery pieces around, pointed them at Phnom Penh, and started a bombardment of destruction and death. They'd fire both cannons, turn them a few degrees and fire again. The rebels kept repeating these actions every few minutes, spreading the firepower in a swath across the city.

As I watched that series of explosions, my thoughts were, "I need to go and see if there is something I can do to help. But there are so many sites. Which one should I go to? The question was answered for me. At the site of the last two shells that struck, a large plume of black smoke began quickly to rise into the sky.

BLACK SUN AND BLOOD RED MOON

As coordinator for the church relief agency and responsible for dealing with the refugees, I left the relative safety of the language school and immediately made my way to the location of the smoke. I needed to survey the damage and see what we could do to help. A horrific sight met me as I arrived at the outskirts of the affected area. The rebels' actions caused a direct hit on the market just as people were returning for the afternoon opening. I saw what I came to see, but it has haunted me ever since.

The market itself was strewn with the effects of the attack. Then my gaze followed the column of black smoke as it rose into the sky. It was behind a row of two-story shops and apartments next to the market. High in the mid-day sky, I saw a black ball. It was the sun. "Wow!" I thought to myself, "the sun is black, just like the prophet foretold, 'The sun shall be turned into darkness, and the moon into blood, before the coming of the great and awesome day of the Lord.'"[2]

I made my way around the market to determine where and how to offer assistance in the face of such unspeakable devastation. There is so much that cannot be told in these pages. It was a dark day indeed.

At the end of that awful day, I returned home, climbed to the top of our building, and attempted to bring my thoughts together and write them down. My eyes once again were drawn to the sky. The moon was a giant full sphere, and deep red, like blood.

On Day 4, God designed a method He would use to speak to His children in all ages. He used Day 4 to communicate something very significant to me there in Phnom Penh. He said, "Jesus is coming for YOU, and this is a sign that you need to be ready. Don't get ready——be ready." This is a sign for all who will be ready for Jesus to return.

SIGNS IN THE HEAVENS—THE SUN, MOON, AND STARS

The lights God placed in the heavens on Day 4 are for signs and seasons! And of course, they have many other functions as well.

Then God said, "Let there be lights in the firmament of the heavens to divide the day from the night; and let them be for signs and seasons, and for days and years; and let them be for lights in the firmament of the heavens to give light on the earth"; and it was so. Then God made two great lights: the greater light to rule the day, and the lesser light to rule the night. He made the stars also. God set them in the firmament of the heavens to give light on the earth, and to rule over the day and over the night, and to divide the light from the darkness. And God saw that it was good. So the evening and the morning were the fourth day."[3]

On Day 4 God chose to make permanent lights in the sky to rule the earth. The Bible information tells us that the evening and the morning of Day 4 comprised a day just like the first, second, and third days. It was neither longer nor shorter. It was not billions of years long, as some say the first day was. The earth rotated on its axis and passed the Source of light. It takes 24 hours for the earth to rotate on its axis. The evening and morning of Day 4 were 24 hours, just as our days are now, and as we saw in Days 1-3.

STORY OR HISTORY?

Remember, Jesus Himself was the Light of Days 1, 2, and 3. The sun and the moon did not yet exist. The Bible tells us they were made on the fourth day. If they were not, it contradicts the narrative of Days 1-3. That makes the whole story just that—a story. There was no attempt to cover or disguise the change from the light which came from Jesus to the light which came from the sun. It would have been easier for God to claim that the sun and moon had been there from the beginning. Such a report would have been easier for skeptics to accept. But God doesn't deal with the easy. He deals with truth. He wants us to know what He was doing, how He made everything. He was building a masterpiece.

Here on Day 4 we are forced to come to grips with our belief in God. Does He exist, or does He not?

Those who want to believe that each day of Creation was some undefined period of millions, or perhaps billions of years, rather than seven literal 24-hour days, often do so with a specific purpose in mind. They do it to allow time for their own conceptualized evolutionary features to develop. You see, if the earth's creation took place in seven literal days, then the millions of years necessary for their evolutionary theories to play out don't exist. They must find a way to fit all this extra time into the week of Creation. But we have already shown Jesus Himself was the Light of Days 1-3. The sun and the moon did not yet exist until Day 4. The days were twenty-four hours long.

It would be easy for a storyteller to describe a time when a make-believe god would be called a great light with the power of the sun. This god-sun might be replaced by a real sun after several days. Indeed, ancient mythology does tell of a sun-god, traveling around the earth, demanding worship by its very presence.

ENOUGH FOR YOU TO BELIEVE—OR NOT

God gave us the truth about Himself. Our salvation is based on that truth—that is, on God Himself. Our salvation depends on the fact that what He says about Himself is true.

If the sun had been in the sky from Day 1, it would have been easier for everyone to say, "That is the sun that was always there. From the very beginning, that was the sun." But the truth is that since Jesus was the Light of Days 1, 2, and 3, there was something that came before the sun, that gave the light to help with the creation of the world. It leads to establish the orderly system of Creation.

Thus, Day 4 strengthens our belief, or lack of it, in God. It gives us the ultimatum of faith. God allows for information that we can evaluate any way we want, and in it, we find the seed of our own salvation or our own destruction. There is an abundance of information available that allows us to make our own decisions as to whether we want to believe or accept God. It is the apple of the Tree of Knowledge of Good and Evil for all who would—or who would not—believe. "And for this reason God will send them strong delusion, that they should believe the lie."[4]

Instead of following God, many in our world choose to find their own way through the pleasures this life might give them. I can tell you that following God has many pleasures. Yet many people choose to follow the pleasures offered by another god—sometimes worshipping pleasure itself. They will turn away from God and be "condemned who did not believe the truth but had pleasure in unrighteousness."[5]

Adam and Eve received this same offer in the Garden. Believe—or not. Eat the fruit of this tree—or not. Worship Creator God, or worship what He has made.

There is no reason to think that this great God who could make the sun and moon in one 24-hour period could not have also made the earth itself in a heartbeat! He didn't require billions of years to do anything. However, He chose to spread the creative process over seven days to establish our weekly cycle.

THE COVER OF DARKNESS

What are other purposes of these lights that God put into the sky? The first thing we read about the lights is they divide the day from the night and divide the light from the darkness. This doesn't make darkness a negative or evil force. It's not something terrible which needs to be overcome by the goodness of light. Dark is not evil, although it is sometimes used to represent evil. It is not of the devil! "He (God) sent darkness and made it dark."[6]

God didn't make anything that was evil. When Moses and Israel reached the Red Sea, with the sea in front of them and the Egyptians behind them, God saved Israel by darkness. He sent darkness to hide them from the chariots.

In 1992 I saw that darkness! My family and I served as missionaries the Republic of the Congo. We had lived in the capital city Brazzaville, for about a year when unrest boiled over in the country formerly known as Zaire. Many missionary families evacuated to safety across the Congo River.

The U.S. Embassy became responsible for all expatriates fleeing from Zaire to the neighboring Congo. Embassy personnel called on the missionaries living in Brazzaville to assist in the effort to house, feed, and offer comfort to these displaced people.

Stations were set up to meet refugees when they came across the river. Temporary lodging housed families while they communicated with their sponsoring organizations located around the world and waited for travel arrangements to be made. Refugees needed a listening ear to provide relief after the trauma of what they had witnessed before their evacuation.

When unrest occurred in Zaire, it soon followed in Congo. The embassy asked if I would carry one of their handheld security radios with me as I continued my work. My task was to observe events and people as I traveled the city streets and walked through the markets. If I saw anything that looked like trouble brewing, I could call them. They would monitor the situation and respond accordingly.

One day, our family was all at home. Our eleven-year-old son came to me and asked, "Dad, why are those two groups of people out in front of our house along the road?"

I said, "I don't know. Let me go take a look."

As we headed out of the house and toward the wall that surrounded the property, he supplied more detail. "Yeah, we were watching, and a motorcycle went through all those people on the other side. When he got to the second group of people and tried to go through them, they stopped him, threw him down, threw his motorcycle down, and started it on fire. Then a car did the same thing. The crowd took the people out of the car and set the car on fire."

I went inside and called the embassy on the radio, "We think there's some trouble here." I described to them what the trouble appeared to be.

They said, "We don't have anybody there who can check it out right now. We'll send somebody over quickly to see what's happening and get back to you. We advise you to stay indoors, in the central part of the house until we can get back to you." We huddled there until finally, the message came on the radio, "Yes, you're right."

We knew we were right because we had already seen what was happening!

"There's nothing we can do about it. However, you shouldn't be there. So, we need you to do two things. First, gather your family and keep them in a safe place. Pack a small suitcase—it has to be a small suitcase—with items that you absolutely need. Each member of the family can bring one small suitcase. We will send the Marines to come get you."

I said, "I understand. We can do that. What's the other thing?"

"There are other expatriates in that part of the city. We don't have anyone else there to warn them. Since you have our radio, we need for you to go on foot and notify the other expatriate families that the Marines are coming for them as well."

During our time in Brazzaville, we had known this might happen someday. As a precaution we had already identified who those families were and where they lived. With the crowds in the street swelling in numbers, the only way to get to those other families was to go either through one of the masses or to go between both crowds.

We didn't have much time. The Marines would arrive in about 15 minutes. Then the voice on the phone said words that didn't need to be uttered. "Be careful."

My wife and I faced a terrible dilemma. Because of the distance from our house to the other homes and from each other, it would take longer than the 15 minutes for me to walk to the half dozen houses we needed to notify. We made the difficult decision that my wife would go to some of them, and I would go to the others. We knew it wouldn't be wise to take our three children through those crowds. They would have to remain at the house at Ground Zero. And so, we prayed. I left the radio with our 11-year-old son. "If anything happens, you get on the radio and call the embassy. Tell them what's going on," I said. "We'll be back as soon as we can."

We opened the gate and stepped out into the chaos. My wife turned to the left, and I went right thru the center of the mob to the other side of the road. It was the strangest thing, however. In our yard I didn't notice anything, but as soon as I opened the gate and went out into the road, a strange green fog seemed to settle over everything.

I had never seen a green fog before. Fog, yes. Green, no. It felt bizarre. I thought to myself, "This is almost like the feeling you have before a tornado, when the atmospheric pressure drops." It did seem as though the atmospheric pressure was lower. But there was something else about it. There was a lower pressure, a green fog, and something else that I've never been able to describe.

Something equally extraordinary occurred as I crossed the street. Neither of the crowds of people facing off with each other seemed to see me at all. As I walked down the road, I noticed a young African man shadowing me, walking a few feet behind me to the right. He kept pace with me, but never got closer. When I turned, he turned.

I finished my task and came back the same way I passed through before. I arrived just in time to watch my wife return from her rounds. As she approached the gate, the mob hurled stones in her direction. Our son opened the gate just enough for her to quickly scurry away from the throng.

The wrong thing to do in a case like this is to run. It shows fear. I deliberately walked across the road between both crowds and entered the gate. My shadow was no longer around. I'm sure an angel walked by my side that day.

Finally, the radio crackled the message we anxiously awaited—the Marines were approaching the house. Grabbing our bags and shutting the door of the house behind us, we went to the gate with our bags. The Marines efficiently ushered us into the vehicle. It was just what you would envision—a truck with armed Marines standing in the back, holding on, watching, surveying, making sure everyone they were responsible for remained safe.

I look back to that event, and I say, God had a reason for what we experienced. Not only that, He definitely protected us. Everything that happens has a purpose and He is leading. A long time after that experience, I decided to check the Bible for any mention of a green fog in Scripture. The closest I could come was as Moses and Israel approached the Red Sea with no place else to go. The Egyptian chariots were coming from behind, ready to attack. God sent darkness. The Egyptians did not attack.[7] Israel walked through the open sea on dry ground.

God uses darkness in many ways. It is not evil. God chose darkness to rule the universe between the places where He installed light "to rule over the day and the night and to divide the light from the darkness and God saw that it was good."[8]

LIGHTS ARE TYPES OF GOD'S COVENANTS

The lights were for more than that, however. They also foreshadowed a type of God's covenants or agreements with man. Covenants are things God said He will do for man—the things He said He will make available for man—for you and for me.

The sun and moon are types of God's covenants. They show how much we can depend on Him to be and do what He says He is and will do. God even compared His covenants to the sun and the moon. "Then this message came to Jeremiah from the Lord: 'This is what the Lord says: If you can break my covenant with the day and the night so that one does not follow the other, only then will my covenant with my servant David be broken. Only then will he no longer have a descendant to reign on his throne. The same is true for my covenant with the Levitical priests who minister before me.'"[9]

God was saying, "My covenants, my promises, are as dependable as are the sun and the moon." Just as we know the sun will be there, in the same way, we are assured His promises are dependable. If you can change the cycle of the day and the night, then you can change His promises. But you cannot change the day and the night.

Since you cannot change the cycle of the day and the night, the evening and the morning, then you know that your salvation cannot be changed by you either. You can only fail to receive the blessing of His covenants if you decide not to accept God's gift of salvation. God will save His people unless we choose not to be a part of that salvation.

GOD IS FAITHFUL TO SAVE US

The lights were also a promise of God's faithfulness, an assurance that God is God and there is no other. This is why Psalm 19 says, "The heavens declare the glory of God." They are a daily fulfillment of the truth that God is the same yesterday, today, and forever, and never changes. He is the One who has always been. He promises to you and me that as sure as the sun and moon are in the sky, as sure as the sun will rise in the morning, God's promises to us are sure. He is the same as He has always been.

The sun and the moon also let you know that Jesus will save you because of who Jesus is and what Jesus did—not because of who you are and what you have or have not done.

God's purpose has not changed and will not change. God made this promise and He sent His Son Jesus. Because Jesus has already lived, died, and risen from the grave, we know that His promise is sure.

We have all heard someone say, "Yes, there are problems today, but the sun will still rise in the morning." The rising sun has always been a promise that God is still in control.

Jesus has done all that needs to be done to save us. It has already been accomplished. We don't have to worry about whether we have become good enough because we know that He is good enough.

A MESSAGE FROM THE MOON

The moon is the lesser light, created to rule the night. But it was more than that. It was the sign of a promise to man that greater light would return. It still says today, Fear not! Remember when the Son is not shining, and there is darkness, the Son will return. There is a new dawn just around the corner.

Sure enough, the sun continues to return every day. The moon returns. As we know the sun will return and the moon will return, we know that Jesus will return. It's not a question of if. It's a question of when.

This part of creation is a continuous reminder about who we are. The lesser light in the heavens reflects the greater light in the sky. We are to be the lesser light, reflecting not our own miserable light—for we are not lights. We are but mere reflectors of light. Still, we were created to reflect the remarkable, perfect light of the mighty Creator of light.

On Day 1, God saw the light which was good, and God separated the light from the darkness. When Jesus comes, He, too will separate or make a clear distinction between the light and the darkness. The brightness of His coming will destroy darkness. Those who reflect His glorious light will be taken to eternity where there will always be only light. When this world ends, God will fulfill all His promises. There won't be any more need for the sun and the moon.

WHEN THE SUN WILL NEVER SHINE

The time will come when He will take back His position, even as the physical Light of the World. The new Jerusalem will have "no need of the sun or of the moon to shine in it, for the glory of God illuminated it. The Lamb is its light."[10]

Jesus is the Lamb of God, slain from the foundation of the earth. The Old Testament foretold that Jesus would become the physical light. "No longer will you need the sun to shine by day, nor the moon to give its light by night, for the Lord your God will be your everlasting light, and your God will be your glory. Your sun will never set; your moon will not go down. For the Lord will be your everlasting light. Your days of mourning will come to an end."[11]

There are many purposes for sunlight that Genesis 1 doesn't mention. To name a few, we know the sun gives heat to the earth. It helps our bodies create and activate Vitamin D. The sun provides the power to drive the atmosphere and runs photosynthesis in plants. It powers all life.

God knew when He created the powerful sun that it would become more than just a wonder in the eyes of some men. Humans would begin to worship the sun and to a lesser degree, the moon, as a god. Nature has long been a god to some. But God calls us to be cautious not to follow those who set the trap of treating nature as god.

LIPSTICK ON A MAKE-BELIEVE PIG

Nature is being used by politicians, scientists, industry and globalists to rebuild the world for them to be top-dogs. We live in the day when nature is one of the most divisive topics in the world. I remember the discussion of global cooling when I was younger. *TIME* magazine published an article titled, "Another Ice Age?"[12] But it didn't happen. Then thirty-four years later, *TIME* ran a cover article entitled, "How to Win the War on Global Warming."[13] Now science is having to go back and alter original data collected years ago. They didn't like the results that showed that the temperature hasn't risen as their prophecies foretold. They have been caught changing weather data from as many as seventy years ago in the upstate New York cities of Auburn, Geneva, and Ithaca.[14]

Global warming is now called climate change. The National Oceanic and Atmospheric Administration (NOAA) frequently makes these adjustments to the raw data. It has never offered a convincing explanation as to why they are necessary. Neither have they shown how exactly their adjusted data provides a more accurate version of the truth than the original data.[15]

This is an example of how we take our eyes off God and make the creation to be our (little g) god focus. Creation becomes more important to us than worshipping the God of Creation.

The Bible warns us, "Take heed, lest you lift your eyes to heaven, and when you see the sun, the moon, and the stars, all the host of heaven, you feel driven to worship them and serve them, which the Lord your God has given to all the peoples under the whole heaven as a heritage."[16]

Don't misunderstand me. God gave man dominion over the whole earth, the air, the water, the animals—and it is our responsibility to care for it and to preserve it. But the present demonstrations based on faulty science and politically correct, but deceptive agendas, are being used to turn us away from worshipping God by worshipping created things.

God knew that it would be tempting for man to worship the creation and to forget the Creator. We will talk more about this idea in Chapter 10. He knew that because of their faithfulness, and their grandeur and power, these great lights in the sky would elicit adoration and passion. Articles from every day of Creation are being worshipped.

This adoration and passion were intended to be directed to God Himself, and not to any created thing. So, He reminds us to worship Him. He points out that worship is the issue of the great controversy between Christ and Satan. The question will be whether we "worship Him who made heaven and earth, the sea and springs of water."[17]

LESSONS ABOUT OUR CHARACTER

We are to be the light of the world while Jesus is gone. This light in us is to cast light in the darkness and allow God to show His faithfulness through us. The sun will not remain our light, but if we are built by Him, our character will last through eternity.

If we are not reflecting the power, glory, and light of God, it is because we have chosen not to. It is because we have put ourselves on the dark side of man instead of allowing God to shine on us and through us.

We are to stand firm in His Word and His faithfulness. We are not to allow the dark times of life to overshadow light He is shining on us. Watch for the signs. Troubles will come, but in the morning, we will see that God is using us for His purpose as we surrender to Him.

Jesus is coming soon. I say this not to engender fear, but to give hope. He is our Creator, our Savior and Redeemer, our High Priest in Heaven. Soon, He will come to take His children home. Are you ready? Are you worshipping Him as Creator? Turn your whole life over to Him today and every day so that when He comes, you will be ready to go home with Him. He is calling you.

Chapter 6

Creation, Day 5

THE SCALE OF BEING

There is a hierarchy, or at least an established cycle through which things were dealt in the Creation. "The measure of divine attention bestowed on any object is proportionate to its rank in the scale of being."[1] We have seen in Days 1-4 of Creation that God created things in a specific order. Day 5 will clarify the reasons for that order and open another window on who we are to God.

The first things God created on Day 5 were fish and the other water creatures. What did the fish need immediately? They needed water. They needed the oxygen that was in the water. Then they needed the plants that grow in the water for food. They needed sunlight that would cause the plants to grow and produce oxygen. All those things were necessary before the fish were made. God waited to create the fish until it was possible for them to survive.

"Then God said, 'Let the waters abound with an abundance of living creatures, and let birds fly above the earth across the face of the firmament of the heavens.' So God created great sea creatures and every living thing that moves, with which the waters abounded, according to their kind, and every winged bird according to its kind. And God saw that it was good. And God blessed them, saying, 'Be fruitful and multiply, and fill the waters in the seas, and let birds multiply on the earth.' So the evening and the morning were the fifth day."[2]

Order of Creation in Genesis 1

God created the heavens and then earth;[3]

He brought Light from darkness in the heavens;[4]

He separated the waters from waters;[5]

He said, Let dry land appear;[6]

He made lights in the heavens;[7]

He created water creatures;[8]

He created the birds in the air;[9]

He made the cattle and other creatures on the earth.[10]

When we get to the second part of Day 6 we will also see man added to the list. We will study this hierarchy in greater depth later in this chapter.

BE FRUITFUL AND MULTIPLY

Before we do that, notice that God gave these living creatures: those in the water, and those that would fly in the air, the ability—and not just the ability, but the command—to be fruitful and to multiply. Why did He do that? As far as we know, this had never been done in the entire inhabited universe. Now we can't prove this because we've not traveled to the rest of the inhabited universe. But there's some indication that it may be so. If it is, then why did God do it with these creatures? Why now? Why on Day 5? Why did He give them the ability to procreate, to work together with creation and continue to produce more of the birds, more of the creatures in the waters? There is no need anywhere else in the universe for plants or animals to reproduce because sin doesn't exist on the other planets. There is no other place where sin abounds.

God knew that this world would be attacked by sin. He knew sin would come and bring death. That would force the need for a continual renewal of life in every species.

We began to study about this already in Day 3. We saw the seeds that God put into all plants. Seeds enable plants and trees to be re-planted and grow more plants and trees. There is no purpose for that if death didn't exist.

God could have decided not to create the earth. He knew what was going to happen. Still, He determined not to stifle the freedom for the man and the woman who would first commit sin, to make their own choices. He does not stifle your free will either. He's not forcing you to do things that you don't want to do. He doesn't use force. Rather He invites. He tells us, "This is what I'd like for you to do, and this is who I would like for you to be. This is what will bring you the greatest happiness." But He will not force you.

Death would come, because turning away from what God wants is turning away from God. When we turn away from God, we decide not to accept the gift of life that only God can give. God put seeds in the plants, and He gave the animals the command, "Be fruitful and multiply."[11] The earth would be maintained long enough to give each person an opportunity to make their own free choice.

When God created the birds and fish and other living creatures, He gave them the process of reproduction. The miracle of birth is God's promise for eternity. God could have used any number of methods to replace the creatures in the sea and birds in the air. He could have repeated His actions the same way He did at Creation. Throughout the life expectancy of this earth, He could have sim-ply continued to make more new creatures. He could have had the angels fly them in from wherever in the universe the fish and bird factory might be established. But instead God created them to re-produce here, on site, in place. There must be a reason.

BIRTH IS THE PROMISE OF CONTINUITY

Birth is the promise of eternity. Birth is the promise of a Savior. Is it possible that the concept of "birth" was created just so Jesus could be born and become God with us—Emmanuel?

As mentioned in Chapter 2, something described as a type is an example of an object that is similar to, or points to another object.[12] When an inventor decides to build a new vacuum cleaner, they will first make a type, or prototype to show potential manufacturers what the device should look like. The prototype points forward to the real vacuum cleaner, which is the antitype. Once the real vacuum is be-ing produced and used in your home, we say "type meets antitype." The real vacuum takes over and represents itself. No one needs to point forward to it. They can say, "That is the vacuum cleaner (the antitype)." There is no more need of the prototype except for his-torical purposes. The inventor will keep the prototype because he put a great deal of work into developing it. It is dear to him.

There is the "type" of the birth of Christ in every creature made by God. Think about it for a moment. These creatures are born be-cause they need to be here for their species to continue until we have no more need of them. That will be when Christ comes to destroy death. Until then God wanted to have a method for these creatures to continue on the earth. They show us that He provided the Savior to rescue us from the sin that causes death.

Keep in mind that not every aspect of literary tools transfer to real life. Once Jesus was born (the antitype), it doesn't mean that other animal births (the type) are no longer needed. They can and do still remind us of the plan of God to point to Jesus. Seeing these animals today reminds us of God's plan to save us.

Every fish that hatches from an egg in the river or lake is the promise of a Savior. Every bird that hatches from an egg, even if that egg is laid in a nest in the highest treetops in the mountains, is a promise of the Savior. A type of the birth of Christ is in every creature made by God. This is true not only in their birth, but in their continued existence.

Adam and Eve were given the promise that the Savior would come. They had no idea how distant in the future it would take for the Savior to come. They didn't know 4,000 years would slip away before Jesus would be born. They only knew that He would come.

They believed when Cain was born, that this gift of God must certainly be the Savior.

Cain was not the Savior. But it helps us to see that to Adam and Eve, birth symbolized the promise of a Deliverer. The birth of every creature is also a promise and a prophecy that the Savior is coming.

DIFFERENT TYPES OF REPRODUCTION

Reproduction is different with plants than with fish, reptiles, and birds. We see the growth of massive numbers of new plants from seeds, usually at a specific time of year. The seeds develop in the plants, and are pollinated by birds, insects, animals or the wind. The plants really have no decisions to make or anything to do with procreation in themselves. Their seeds simply develop, and then fall at the appropriate time. Some even require fire to break the seeds from their pods to prepare them to germinate. The plants never have anything to do with them again.

The birth of animals also usually comes at a specific time of the year. They propagate en masse. These creatures are birthed by the millions. There are places on the earth you can go and see more than anyone can count. Their continued propagation was designed to be that way.

It was not made to be a matter of loving parents fulfilling their passion for one another and procreating a new soul. The mating is done by instinct. The creatures can't help it. They can't make it happen, they can't prevent it from happening. With few exceptions, at certain times of the year, the creatures all experience the drive to get together and mate. It's not something they have any control over. This is one way to recognize that animals hold a lower rank on the scale of being than we do. The smaller creatures usually reproduce in greater numbers than the larger animals.

Why did God make them that way? Perhaps they were created to show that the birth of Jesus was to serve or be available to the masses. They make it clear in massive frequency that a Savior would be born. God wanted us to see over and over, everywhere we looked, the promise that Jesus the Savior is coming.

MAN'S REPRODUCTION IS DIFFERENT

God made man and woman one at a time. Perhaps this was done in such a way that the birth of Jesus would also be seen to be uniquely for you and for me. He was not born for the masses, but for us personally.

Or perhaps the salvation brought by His birth, His life, His death, and His resurrection or rebirth can be seen to be uniquely the antitype of our rebirth when we are born again.

God designed human procreation to be none of the things that the other creatures of the earth experience. Rather, it was to be personal, loving, relational, controlled, planned, thought out, and intentional.

This also shows how we have fallen since Creation. Today many babies are born by accident, due to uncommitted personal gratification, or drug induced conception. In 2014, this led to the murder of nineteen percent of babies in the United States before they are born.[13]

It is this very creation of man by God and our gift of intentional procreation that continues the hierarchy that God made. The fact that God made man to be unique reveals our level in the hierarchy. While we won't get to the creation of mankind until Chapter 7, it is easy now to see that God worked differently when He created man and woman.

BIRTH MADE TO DRAW PARENTS TO GOD

Another reason why birth was created was to draw us closer to God Himself. "After the birth of his first son, Enoch reached a higher experience. He was brought into a closer relationship with God and realized more fully his own obligations and responsibility as a son of God."[14] Every one of us who has become a parent can see how birth draws us back to God. As we look at the tiny wrinkled form we see the miracle of what He has done for us, in us, and through us. We are allowed to create a new soul in our image. We can recognize the miracle and think, "Oh, what a mighty God we have." Birth draws us back to God.

At least for a short time, we worship the Creator God. We recognize our dependence on Him and our importance in God's eyes.

ANOTHER DIFFERENCE BETWEEN US AND OTHER ANIMALS

How can we be sure that there is a difference between man and all these other creatures? Is there a difference? This very topic is being debated around the world today. We will take a closer look at this in Chapter 8.

But here's one way the difference is clear. God created massive numbers of the birds. He asks, "What is the price of two sparrows—one copper coin? But not a single sparrow can fall to the ground without your Father knowing it. And the very hairs on your head are all numbered. So don't be afraid; you are more valuable to God than a whole flock of sparrows."[15]

There are billions of sparrows around the world, not to count all the other animals and birds. God sees every one that falls. We are of greater value to God than many sparrows. Matthew shows us a lesson we are to learn from the masses of birds. Man was made only one day after sparrows, but our value is much greater than many of them.

ANOTHER WAY TO DETERMINE VALUE AND RANK

As was stated before, we can determine the rank God places on everything by the attention He bestows on it in Creation. Part of this attention is determined by what was done before a creature was made, showing how much care God took to be ready for it to live.

Everything was made to have a place and a rank. There are some things God sees as being more important than other things that He made. The little brown sparrow is watched over by God. The flowers of the field, the grass that carpets the earth share the notice and care of our heavenly Father. He cares even more for man who is the image and glory of God Himself.

Genesis 1 establishes the scale. God did it on purpose. We see that there was dirt, water, atmosphere, dry ground, plants, food, birds, fish, and then mammals, cattle, deer, and finally man. The rank is obvious. This was not established by man, but by God Himself.

You are positioned in the highest place in the scale of being of any created object. As such it is to you who God bestows the highest measure of His divine attention. Do you feel unworthy? Do you feel unlovable or unloved? Do you feel ineffective, untalented, unimportant, or any of those un- words? You are none of those things. There is nothing in this world that is more important to God than you. He loves you more than anything else on Earth that He created.

RE-ORDERING THE HIERARCHY

In the last days before Jesus returns, we see this becoming an important issue for humanity. Man is re-ordering the hierarchy that God places on creation. Have you seen it? This re-ordering places the creation at the pinnacle and not God. "Fear God and give glory to him for the hour of his judgment has come and worship him who made heaven and earth, the sea and the springs of waters."[16]

These words are recorded in Revelation as a challenge to us. He warns us that in these times that we live, the world will be trying to take God out of the picture and raise the creation to be more important than Him, more important than man.

Revelation 14:7 is a call in the last days not to fall sucker to the movements of the day that place the creation above man in importance. There is nothing on earth more important to God than mankind. Nothing more important than you. The light is not more important than you. The air, the dirt, the water, and the atmosphere are not more important than you. Not the birds, not the fish, not the animals. There is nothing more important to God than you.

Man is of supreme importance to the Creator. He gave all these other things for us to take care of and have dominion over.

This doesn't mean that we may abuse our earth and what is in it. It doesn't mean that we can be unconcerned about the condition of animals or the environment. We are the caretakers, and we will become responsible to God for how we take care of His Creation. However, the earth is not our mother! It is our temporary home!

DON'T WORRY!

Since you are one of God's highest-ranking beings in the order of earth's creation, you have no need to fear. So, don't worry! Luke brings us back to the relationship between these lower rank objects and mankind. Jesus gave this descriptive lesson about the way God sees us in comparison to the rest of Creation.

"I tell you not to worry about everyday life—whether you have enough food to eat or enough clothes to wear. For life is more than food, and your body more than clothing. Look at the ravens. They don't plant or harvest or store food in barns, for God feeds them. And you are far more valuable to him than any birds! Can all your worries add a single moment to your life? And if worry can't accomplish a little thing like that, what's the use of worrying over bigger things?

"Look at the lilies and how they grow. They don't work or make their clothing, yet Solomon in all his glory was not dressed as beautifully as they are. And if God cares so wonderfully for flowers that are here today and thrown into the fire tomorrow, he will certainly care for you. Why do you have so little faith?

"And don't be concerned about what to eat and what to drink. Don't worry about such things. These things dominate the thoughts of unbelievers all over the world, but your Father already knows your needs. Seek the Kingdom of God above all else, and he will give you everything you need."[17]

God has taught us this lesson when He established our place in His scale of being. He is still in control. Rather than worrying we must seek the kingdom of God and His righteousness and all these things shall be added unto us. "Don't be afraid, little flock. For it gives your Father great happiness to give you the Kingdom."[18]

Through Luke, God reinforces the scale of being. In the end He reminds us He is coming to give us the kingdom. All of this is a promise of the kingdom. He did it on purpose. Light, atmosphere, water, dry ground, plants, food, animals, birds, fish, mammals, and finally man. When we see that this is true, then "Sell your possessions and give to those in need. This will store up treasure for you in heaven! And the purses of heaven never get old or develop holes. Your treasure will be safe; no thief can steal it and no moth can destroy it."[19] "Seek the Kingdom of God above all else."[20] "Wherever your treasure is, there the desires of your heart will also be."[21]

This is the lesson of all these creatures that God made on Day 5. Luke 12 repeats the phrases—do not worry, do not be anxious, do not seek, do not fear. All together they tie this into a whole thought that verifies rank in the scale of being. We recognize and accept we are the highest object of the creation of God. We are in His image. This leads us to surrender to Him and to accept that He will take care of us. He will provide for us. Don't worry or fear, but seek Him first and seek His great provision.

These promises don't mean there may not be difficult times. It doesn't mean we'll all have everything we want. It doesn't mean we can always see what's coming before us. God gives us these reminders to show us little things along the way to help us see He is indeed providing for us.

We are at the pinnacle of the rank of the scale of being. To play our appropriate role we must accept our place there and treat the rest of Creation with respect and honor as God instructed us to do.

God gave mankind a role in procreation. He made us to be higher than the animals. They were not created to be in His image. Since you are higher than the animals, live like it!

Recognize that last day pseudo-science seeks to turn man from worshipping God by elevating the creation above man and even above God. Don't fall for it!

God is still in control. His great plan will be fulfilled, and He will take care of you. Don't worry. Be happy!

Chapter 7

Creation, Day 6A

THE HIGHEST ORDER

When God puts something in Scripture He wants us to notice, He repeats it. We saw this modeled earlier on Day 1 and Day 3. We find the same thing in the description of Day 5. Now it is repeated in Day 6. "Then God said, 'Let the earth bring forth the living creature according to its kind: cattle and creeping thing and beast of the earth, each according to its kind,' and it was so. And God made the beast of the earth according to its kind, cattle according to its kind, and everything that creeps on the earth according to its kind. And God saw that it was good."[1] In just two verses, we find the phrase "according to its kind" repeated five times. Since God put this clue here for us, we need to understand why this is so important.

Then God said, "Let Us make man in Our image, according to Our likeness; let them have dominion over the fish of the sea, over the birds of the air, and over the cattle, over all the earth and over every creeping thing that creeps on the earth."[2]

Evolutionists continue to research our origins. Their theory is that over millions and millions of years, we have evolved into who we are today. But God says He made all the animals "after their kind." He made them in a specific way, and they continue to reproduce after their kind.

There's no question for those who believe in God that He was faithful to His word and followed it with action. After all, He is the One who made everything to begin with. He also revealed the method by which life would continue and how the species would survive. It is not by evolution and natural selection.

The animals described by the generic term cattle were created on Day 6, in the same way as the birds and fish were created on Day 5. They were designed to reproduce after their kind. When the Bible speaks of cattle, it is not only referring to what we call cattle, but to all the larger, four-footed animals that populate our earth.

As I have already mentioned, I grew up in Western Kansas. Although wheat is a major crop, we also raise a lot of cattle. Any Kansas boy knows what cattle are when he sees them. You see herds and herds of them grazing in the pastures. Depending on the wind, you often know they are there long before you see them.

But when God said, "Let Us make man in Our image, according to Our likeness," it was something different. Animals were not made in His image. We were.

GOD FINISHED CREATING "OBJECTS" ON DAY 6

Day 6 is the last day God created objects. With all the pieces in place as a foundation, God began creating institutions. For example, He instituted marriage on Day 6. On Day 7, He declared the institution of the Sabbath. We will look more at what these foundations are in Chapters 9 and 10. But marriage and the Sabbath are not things. They are ideas. They are principles. God also created or described on Day 6 what we might call lifestyle patterns, such as diet. We can see what He intended for us.

ANOTHER STORY OF BEGINNINGS

Let's look at another story of beginnings. There are two major narratives of beginnings with which we are all familiar. The first is the Creation story given in Genesis. The other is called evolution.

The order of evolution loosely follows the order of the development given by God's process of creation. In other words, life developed from the simplest form to the most complex. This theory follows the Creation narrative. In both, the foundations were laid first. After that, the least developed and lower-ranked creatures were made. At that point, the story of evolution pivots far away from Genesis in many ways.

The evolutionary theorist, whose beliefs are almost like a religion,[3] says that to move from simple to complex, modifications, or mutations occur in their DNA changing living things from one form, or species, to the next. (We will look at genetic entropy in Chapter 13.) If evolution were true, then because of the great volume of scientific research, there would be evidence.

Such an explanation turns away from God's description of how things were made. He made the simple, then moved on to create the more complex, and finished with the creation of the most complex. God made them the way He intended them to be. As part of that, He made the offspring of each after their kind.

FOSSILS TELL THE TRUE STORY

What is the evidence of evolution? Do fossils confirm evolution? It is true many fossils are found the world over in many locations. They often show different animals than those in existence today. The uniqueness of these remains is interesting to study. When we examine fossils, the evidence of the skeletal remains shows us these animals certainly came from somewhere, and they did exist.

Some Christians try to deny these creatures ever existed. The evidence is clear. We cannot deny scientific proof that is so easily observable. We can question different things about the evidence, and we will look at some of that later. We should not refute science just because it may seem not to agree with our belief. Scientific discoveries deserve consideration and require correct interpretation. Remember, true science comes from the same source as true biblical knowledge.

Evolution tells us that mankind started as a simple life form and through millions of years evolved into what we are today. We have all seen the series of pictures that show a chimpanzee, several other species, and finally, people like us.

You might expect with all the completed research, and the number of fossils that have been found, you would be able to see many examples of this transition from one species to another. The problem is that there are no fossils that display this transition process in any species, on any continent, at any time. The fossil evidence simply does not exist. Science has proven that the transitional process of evolution never happened.

However, some scientists report this lack of fossil record differently. They say they just haven't found the transitional fossils that show us the change from one species to another. They give misleading examples, calling them evidence of species transitions.[4] Their examples come from faulty misapplication of the presupposition that a species can change to become another species. It is circular thinking. It sounds like this: "We believe that a species can change to become another species. Therefore, when we find fossils that have some differences as well as some similarities, we call that proof that the first changed to become the second. This proves we were right in our thesis." But it doesn't. This evidence actually comes closer to proving that they were created after their kind than it proves evolution.

The fossil evidence tells a much different story than the anti-God theory, and evolution is exactly that. It really is not a theory of the transitions of humans and development of man through an evolutionary process. It simply says, we can't prove how it happened, but we know it wasn't God.

THE MISSING LINK IS STILL MISSING

For example, let's look at a scientific discovery called Piltdown man discovered in Sussex, England between 1908 to 1912. It is a fossil of a skull. How did it take four years to discover a single skull? Amateur archeologist Charles Dawson discovered a human-like skull fragment in gravel beds near Piltdown in Sussex England. Archeologists gathered the fragments discovered in various places during that period and used them to piece together what they called a human skull. It was named for the area where the fragments were discovered.[5]

The announcement to the public was that scientists had discovered a transitional form of man.

Much was made of this link to man's evolution. For 41 years, scientists insisted that Piltdown provided evidence that man evolved over time. In 1953, the Natural History Museum announced the Piltdown discovery was a hoax. They proved what everybody should have already known. Taking fragments of bones from different places over a four-year period and putting them together in the way you want them to look does not prove a theory. Not only were the pieces taken from at least two different locations, but they were from more than one kind of creature! Four decades after the discovery, science had to confess the error.

The issue here is not simply that errors were made. It is also that two generations who were taught evolution on the basis of this hoax were raised believing a lie. On this foundation, many chose to turn away from a belief in God. It is unlikely, too that the evidence of the error was as widely disseminated as the remarkable discovery had been. It is admirable that the Royal Society Open Science continues to work to understand and to publish, as late as 2016, on efforts to expose the fraud.

Another example of this is a skull named Ida. It was presented as the "link between us and the rest of the animal kingdom."[6] As it turns out, it may have been related to the lemur or the loris. Whatever it was, it has gone extinct. Every time scientists say that they found evidence of a transitional fossil, it has been proven by science to be wrong. There are no scientific discoveries that show the process of evolution taking place. There are no irrefutable examples of transition from one species to another.

Consider as well that if all creatures evolved from one form to another, we should see such creatures all around us. Chimpanzee-like creatures are said to have begun to change into the next species of bipods—two-footed creatures—from two and a half to seven million years ago.[7] According to the theory, those bipods changed into the next species of bipods, and then the next species. You get the picture. Finally, all this changing got to us.

Why do we still have chimpanzees today? Why didn't they all change and transform? The fact that we still have chimpanzees makes it obvious that not every individual of a species made the same changes. That being the case, there should be animals and humans in each of these "former stages" as the process continued. It should be ongoing today. Since we obviously do have chimpanzees, why don't we also see all, or at least some of the other stages of transition still living in the same jungles? Chimpanzees would have continued over time to transition through the evolutionary process to humanity. But it isn't happening.

This creates a dilemma for the evolutionist. While we have many chimpanzees and many humans, we have no examples of any transitional beings.

Today, we cannot find examples of other forms of man. The evidence disproves scientific claims. True science is observable and/ or can be reproduced. In this case, science proves that evolution did not happen.

The only examples of transitions between one form (species) of creature and another are those that man has built from unrelated fragments of bone fossil.

Since no examples exist showing any transition of any species, on any continent, at any time, it is time for science to recognize that the theory isn't true.

"And God said, 'Let the land produce living creatures according to their kinds: the livestock, the creatures that move along the ground, and the wild animals, each according to its kind.'"[8] Thousands of years before evolution was ever considered, Scripture described for us the progression of the species. Each would come from its own kind, not from some other species. God placed this clearly in Genesis, so we would know in clear words exactly where we come from.

LET IT BE, MAKE IT HAPPEN

When God said to let something happen, it meant that He would make it happen. He would do it. Nothing happens without God letting it or making it happen. God said, "Let us make man."

Then He made it happen. He said, "Let there be light," and there was light. He said, "Let the waters separate from the waters," and the waters were separated from the waters. He said, "Let there be birds and fish," and then He made them.

He let it happen. For each thing that He let happen, the Bible also says, "God made the firmament" and "God made the two great lights" and "God made the great whales, and every living creature that moves."

By letting things come into being, God somehow removed the physical barriers that may have existed to prevent these things from existing and surviving. He changed the laws of metaphysics and let these things become, simply by allowing them to happen at His command.

IN GOD'S IMAGE

God made man in His image. Nothing in creation was so contrary to the pre-existing conditions of the universe than for something to be made in the image of God. Think about it. If you were God, would you make something in your image, especially something that you knew would try to overpower you and take your place? But God did. He removed that barrier as well.

It is here that we begin to see the significance of God's love for us. This language of Scripture assures us that God made each of us to be the image of Him. He intentionally made us to be a selfie of Himself, so that others would see Him when they look at us. That reveals the great esteem He has for us. It also reminds us of the responsibility we have to Him, that we will not betray His love for us.

Consider the craftsman mentor—such as the blacksmith or cabinet maker during the nineteenth century. In the frontier towns of the developing country, usually only one person could provide the specialized services of creating items out of metal and wood. Generally, only one specialist in each trade was needed, unless the town was growing rapidly. When each craftsman neared retirement, he trained a younger person to carry on with the craft and trade. Someone would be ready to replace him. If someone could replace God, then making someone in His image might invoke the concept of training mini-gods. God didn't need someone to replace Him. He is eternal. Besides, no one could.

God took great risk for Himself and the entire universe to create another being in His own image and after His likeness. The fact that so many people throughout earth's history have tried to make themselves to be God proves the danger of His choice to create humanity in His image. But God's great desire to have this being to love and to be loved by, overruled all the barriers. God was willing, perhaps even compelled by His love for us, to make beings in His own image.

"Then God said, 'Let us make mankind in our image, in our likeness, so that they may rule over the fish in the sea and the birds in the sky, over the livestock and all the wild animals, and over all the creatures that move along the ground.'"[9]

God made man and gave him some of the abilities and powers that He possessed. What does it mean, and what does it not mean that we were made in the image of God?

Man was made in the likeness of God. He may look something like God looks. Making man in God's own image does not mean making him to be God. Recognize the distinction. It does not mean that man was to portray every character trait or ability that God possesses. There are some attributes of God that were freely created into Adam and Eve. But some traits would have made man too much like God. These were not entrusted to us.

We can understand it better by seeing the same concept when Adam's third son was born. "This is the written account of Adam's family line. When God created mankind, He made them in the likeness of God. He created them male and female...When Adam had lived 130 years, he had a son in his own likeness, in his own image..."[10] Seth was born in the image of Adam, after his likeness. Was Seth Adam? No, he wasn't. Was Seth the same as Adam in every way? No. Made in the image and the likeness of Adam, Seth was similar to Adam in some ways.

Recognize that the female of our species was also made in the likeness of God.[11] For us, when we have children, whether they are boys or girls, they are in our likeness, and in our own image.

God made us in His image, but He did not make us God. Not only are we not God, but neither are the angels! When the angel, Lucifer, decided he wanted to be God,[12] the trouble in our universe began! God knew that man would sin, so some traits God could not entrust to us. We will discuss this more in Chapter 12.

This was not done to limit man. It only limits the damage he could do in the universe, had he been allowed to exist in certain dimensions where God freely moves. There is no one like the Lord our God.[13] This is an important distinction because some people say God made us like Himself. We are not God. There are attributes that belong to God that do not belong to man.

1. God Alone Is Immortal.

This attribute of God is often misunderstood when we consider man's likeness to Him. Today, people talk about our immortal or eternal spirit or soul. But the Bible tells us about "God, the blessed and only Ruler, the King of Kings and Lord of Lords who alone is immortal."[14]

There are some things that man was never intended to be. However, God did intend for man to live forever. That is clear by the fact that the earliest generations lived to be more than nine hundred years old, even after sin entered and its curse fell upon the earth.[15]

Sin changed what God had intended. Many accuse God of allowing bad things to happen. We want Him to prevent others from doing evil. Still, if we were prevented from our own brand of mischief, we would be just as angry at Him. Sin is our choice. Its consequences are our reward.

The reward of Adam's sin was the curse on the earth. It was in this curse that man's mortality was first mentioned. As will be discussed in Chapter 12, man's death was not the curse that resulted from their sin as much as a natural consequence of it. "'By the sweat of your brow you will eat your food until you return to the ground since from it you were taken; for dust you are and to dust you will return.'"[16] Death was not God's original plan for man.

There are only two people about whom the Bible records that they did not die. They are Enoch, who walked with God[17] and Elijah, who was taken to Heaven in a fiery chariot.[18] With the exception of these good people and those of us who will still be living when Jesus returns to take the righteous to a heavenly reward, every other living soul will die. This is the fate of all but two people from the time of Creation until Jesus returns again.

This is not only a physical death, but the death of the soul as well. "Behold, all souls are Mine; The soul of the father as well as the soul of the son is Mine; The soul who sins shall die."[19] That is because sin destroys the image of God in us and separates us from the Life-Giver.

WHEN IS IMMORTALITY GIVEN?

The Bible gives us both a description and an explanation of the event. It speaks clearly about those who "believe that Jesus died and rose again…For the Lord himself will come down from heaven, with a loud command, with the voice of the archangel, and with the trumpet call of God, and the dead in Christ will rise first. After that, we who are still alive and are left will be caught up together with them in the clouds to meet the Lord in the air. And so we will be with the Lord forever."[20]

This speaks of two groups of people. The first group is the people who died, believing in Jesus. It is not those who had a shallow belief. These are truly "in Christ." They have surrendered to Jesus and accepted His gift of being recreated. They are the people who have died since the beginning of time, believing in God[21] and surrendered to Him. They will be raised from their peaceful rest by the voice of Jesus and the trumpet call of God. It will be just like when Jesus called His friend Lazarus to life.[22]

The other group of people with the same belief in Jesus as the first, are those who will still be alive when Jesus comes. They will have endured the struggles of persecution and trials. They will see Jesus coming to Earth in clouds of angels and calling their names. They will rise to meet Jesus in the air after the parade of those who will be resurrected. Together, these two groups will journey with Jesus to an eternal reward of immortality, joy, and peace. By this time, anyone who does not have this deep faith will not remain alive on earth.[23]

The change from being mortal to being immortal will occur at that moment when Jesus calls the rescued back to life. "It will happen in a moment, in the blink of an eye, when the last trumpet is blown. For when the trumpet sounds, those who have died will be raised to live forever. And we who are living will be transformed. For our dying bodies must be transformed into bodies that will never die; our mortal bodies must be transformed into immortal bodies.[24, 25]

In Heaven and the New Earth[26, 27] our immortality will be maintained by eating fruit from the Tree of Life[28, 29] Our immortality is not the same as is God's. We will still depend on God's provision for eternal life, even if it comes from eating from the Tree of Life. God does not depend on anyone or anything. In that way, we will not be like God even when we are in heaven.

2. God is All-Powerful—Omnipotent.

God's great power is not made of muscles and fortitude. It is the same power He used in Creation. He said, "My hands have made both heaven and earth; they and everything in them are mine. I, the Lord have spoken."[30] He speaks, and things appear from nothing. He imagines and then watches His imagination become real. There is no indication in Scripture that we will have this kind of power in heaven, although we have shared in the power of procreation in this life.

3. God is All-Knowing—Omniscient.

Some things were never intended for man. For instance, God never intended that we know good and evil. Can we be sure He didn't want us to know that? There was a way man could know good and evil, but God told them not to eat of the tree that would give them this knowledge.[31] He didn't want us to know good and evil.

The devil told Adam and Eve the truth when he said, "God knows that your eyes will be opened, as soon as you eat of it, and you will be like God, knowing good and evil."[32]

Was that true? Some people say the devil is a liar. Yes, he is. But when it serves his purposes, he can tell the truth too. This was the truth. The moment Adam and Eve bit into the fruit's flesh, their eyes were opened, and they did become like God in that they knew good and evil. How do we know?

We know it was true, because "the Lord God said, 'Look, the human beings have become like one of Us, knowing both good and evil.'"[33]

That is exactly what the devil told them would happen to them if they ate of the tree. In this small way, possibly the worst way, we became like God, particularly knowing evil.

In one sense, Adam and Eve had never known good. When you have only one thing with nothing to compare it to, you may not realize what you have. When your life is only one way, you can't make a comparison. Have you ever heard someone say, "When I was growing up, my family had nothing. I never knew we were poor until I got older and saw the rich."

Adam and Eve didn't know there was anything such as good until they saw evil. The place they recognized evil was within themselves. God was able to know good and evil without partaking of the evil. But for mankind, that was impossible. They learned of evil by sinning.

God Himself said that man had become like Him in that terrible, awful knowing. We now know good, and we know evil. How is that serving us? Look at the cost to Adam and Eve. Look at the cost to every human being who has ever lived. God knew what good was, and He knew what evil was. Until he sinned, man only knew good. After he sinned, he began to know what evil was. Why? Because he felt it in his heart. God never intended that we feel the pain of evil. He wanted to keep that from us.

The only way the Bible specifically tells us that we are like God is that we know good and evil. It is certain that we do not know the extent of either good or evil. But we can know the difference between the two. We can distinguish between what is good and what is not good. God helps us to be able to do that by giving us His Word, the Bible, and by giving us the still, small voice of His Holy Spirit. We can hear that voice tell us, "You should probably stay away from that."

Look at the cost of becoming like God when we were not prepared to be like God. "So the Lord God banished them from the Garden of Eden, and He sent Adam out to cultivate the ground from which he had been made."[34] Living forever in sin would have caused sin to last forever. God could not allow that to happen. It would destroy the universe the same way it has destroyed this planet. That is why He made the time box to contain evil and sin.

Man traded the privilege of living forever, to receive a life of labor, followed by certain death. This is the kind of trade we often make with God when we think we have a better way to do things than the way He tells us.

4. God is Present Everywhere—Omnipresent.

God is everywhere at all times. This is as much beyond our ability to comprehend as are His immortality, His power, and His knowledge. King David said to God, "I can never escape from your Spirit! I can never get away from your presence! If I go up to heaven, you are there; If I go down to the grave, you are there. If I ride the wings of the morning, if I dwell by the farthest oceans, even there your hand will guide me, and your strength will support me."[35]

God's ability to be wherever and whenever David was, tells us He has the same ability with us and everyone else.

Jesus gave up the ability to be everywhere at once when He became human and was born on earth. He will forever be limited to be in one place at a time. We are told to "have the same mindset as Christ Jesus: Who, being in very nature God, did not consider equality with God something to be used to his own advantage… being made in human likeness and being found in appearance as a man, he humbled himself by becoming obedient to death—even death on a cross!"[36]

After His death and resurrection, Jesus was taken back to Heaven and given a new glorious body, but without omnipresence. When He comes again, "He will take our weak mortal bodies and change them into glorious bodies like his own, using the same power with which he will bring everything under his control."[37]

We will never be God. Nor should we aspire to be. We should never wish to be gods, for to wish that is the very sin of Lucifer in heaven that started the sin problem. Lucifer said, "I will climb to the highest heavens and be like the Most High."[38]

The desire to be fully what God is promotes personal aggrandizement that leads to pride that comes before a fall.[39]

WHAT ABOUT THE IMAGE OF GOD IN YOU?

You are exactly what God wanted. If you give God permission, He will build you back to His image. You and I can't make ourselves into God's image. We can't rebuild ourselves to be in the image of God, after His likeness. That is not our job. That is God's job. Our role is to surrender to Him and ask Him to forgive us for our sins. Then God cleanses us from everything that is not good.

God will not take sinful things from our hearts until we are truly ready for them to be gone. He will not force you to allow Him to rebuild His image in you. God will not force you to surrender. When we are ready to accept Him and His wonderful gifts and allow Him to remove the evil from our hearts, He is more than happy to do the work in us. We call it sanctification, and it is the work that God will do in you for a lifetime.

If God planned you, made you, and loves you, He will not abandon you or turn away from you. He will always be with you.[40]

Do not look at others and say, "If only I were like that." There are some things you are not intended to be or to have. If you were intended to be them or have them, God would have made you to be those things. Find joy in the way He is leading your life. But be sure it is He who is leading you.

God did not take chances with your salvation. He did not send an angel. He came Himself. Only He could pay the price for your sins and mine. But no one else can accept this great salvation for you. Only you can do that. What will you do? Will you accept the great salvation God is waiting to offer? It is your choice.

Chapter 8

Creation, Day 6B

MANKIND IS IN CONTROL

For man to be made in God's image and to be remade in His image are both the work of God. We had absolutely nothing to do with the original work God did. Neither do we perform the new work that must be done in each of us. The difference is, we have ability and the responsibility to control what God does in remaking us. Each of the three steps must be accomplished. Each requires our surrender.

Step 1: reMaster

That is, make Jesus the Master of your life. The Bible describes it this way. "If you openly declare that Jesus is Lord and believe in your heart that God raised Him from the dead, you will be saved. For it is by believing in your heart that you are made right with God and it is by openly declaring your faith that you are saved."[1] The word Lord means Master. This is a surrender of our will and our life to Jesus.

We recognize that we can't be good enough or always do the right things. But we believe that Jesus the Creator has the power to do that in us. He proved it by rising to life after dying to pay for our sins. Since He won't force us to do anything we don't choose to do, we ask Him to be Master of our lives, instead of being our own master. Doing this puts us in right relation with God. Before any change is made in our conduct, He accepts us.

Step 2: reImage

God takes over from here. He will do the rest, as long as we don't choose to take back control of our lives. He will never force us to do His will. God will renew His image in each one who has been reMastered. "For God knew His people in advance, and He chose them to become like His Son…"[2] God will do the work of making us in the image of Jesus. He, through the Holy Spirit, is the only One who can do this work. He made man in His image at Creation. He can remake us in His image if we allow Him to do so.

Step 3: rePurpose

When the image of God is being renewed in us, we will want to be what He leads us to be and do His will. We see that God did not call us to Himself to sit idly by while others are lost. Our lives have a divine purpose. "And we know that God causes everything to work together for the good of those who love God and are called according to his purpose for them."[3]

We know what His purpose is. "Your Father in heaven is not willing that any of these little ones should perish."[4] In the same way He wants us to be ready for heaven, He wants all His other sons and daughters to be ready, too. This is the reason Jesus gave explicit final instructions to His disciples before He left them to return to Heaven after His resurrection. "Go and make disciples of all the nations, baptizing them in the name of the Father and the Son and the Holy Spirit. Teach these new disciples to obey all the commands I have given you."[5] The rePurposed person will be about his Master's business.

Day 6 is the only part of Creation that some people are interested in. It is the day that many consider the most important because we were made then. The fact that there were at least two major segments of the creative process gives us another indication of what was noted concerning the rank in the scale of being. By our definition, the latter portion is higher on this scale.

LEARN FROM THE PATTERNS

We've already noticed one way God emphasizes important concepts He wants us to understand. He uses words or phrases multiple times in a short space when He wants us to focus our attention on a particular idea or concept. At times, He repeats common words that we might pass over, without realizing there is something significant hidden in the message. We don't want to fall into that trap. We looked at this on Day 3 in our study of the seeds.

We find the same pattern in Day 6 as well. Here again, simple words and phrases are used multiple times. God is saying, "Look at this! I don't want you to miss this great lesson!

What is the common word or phrase God used when He was talking about the creation of the animals? In Genesis 1:24-25, God talks about creating the animals and repeats the phrase "according to its kind." He uses it five times in those two verses. What is so significant here that God emphasizes this point?

It is God's signal to us that He wants to teach us something out of the ordinary. With a simple reading, we can grasp the surface concept, but there is a great deal more to learn.

Why are we studying Creation? "Worship Him who made heaven and earth the sea and springs of water."[6] Today, just before Jesus returns, false science, which includes evolution, declares the message that over hundreds of millions of years all life came from lower forms of life.

But Creator God says, "I made them according to their kind." He didn't make them from some animal that may or may not have been somewhat like them. It is true that when God created them, He made things that never existed before. They were originals. Their offspring would be made after their own kind. He didn't change them over years, centuries, or millennia to become different species.

The fact that certain creature fossils only occur in certain levels of sediment deposits does not prove that they did or didn't exist before or after other species. In fact, it is evidence that they all died in the cataclysmic event we call the Flood. They were deposited in their layers according to the currents of the tumultuous deluge.

Revelation returns our focus to worshipping God as Creator. Life didn't evolve. Genesis shows us it came from the same kind of life that came before it. God repeated this mantra five times to drive the point home to us in these last days.

This message is in these verses for this day in which we live. He placed this explanation in the Bible thousands of years before it would become an issue so that we would have the truth to counter evolution. This was intentional.

HISTORY OF CHANGING IDEAS

In the seventeenth century, Anglican archbishop of Armagh in Northern Ireland set the date of Creation at October 23, 4004, BC.[7] It would be interesting to see his research.

Historically, Creation was generally accepted as the developmental method of life on Earth. Even through the nineteenth century, belief in Creation was shared by scientists.[8] But there was a movement afoot. Erasmus Darwin, an amateur scientist and the grandfather of Charles Darwin, published his ideas and poems about evolution in 1794-1796.[9] It wasn't until around 1801 that Jean Baptiste

Lamarck proposed a "full-blown theory of evolution."[10] At that time it seems science began to take the concept of evolution seriously.

Most people who have any interest in science or Creation/evolution know the name of Charles Darwin. In 1859, his book, *Origin of Species* was published and is said to have sold out in one day.[11] Today, the study of Creation is an issue. Because of the work of these men, Charles Darwin, and a myriad of others in our day, the theory of evolution receives strong acceptance in our world. Science continues to advance these ideas, even in the face of strong evidence that their conclusions have been flawed. This is not to say scientists as a whole lack in intelligence, or that they intentionally lead their research to faulty conclusions. Some may be plagued by the failures of others in the past. Others have failed to re-evaluate some of their foundational research principles and assumptions on the basis of improved methods and technology. There has been enough fraud in the ranks to bring the whole concept of evolution into disrepute, but there is an even larger reason to see the ideas as incorrect.

The reason is science itself. It is understandable that most researchers have been honestly following in the steps of those who preceded them. But they can become entrenched in supporting the status quo beliefs that have brought them to their lofty heights of academia and research institutions.

Still, if someone is truly a scientist, they will follow the science where it leads. They won't dismiss new science just because it reveals different truths than they had previously believed. It has long been believed that during the Enlightenment, science and the church were in conflict over their differences in perspective. Today, this "conflict thesis" seems to have fallen out of acceptance among contemporary historians of science.[12] This sounds like revisionist history. Still, today, new science could be coming into conflict with old science.

Ben Stein, an accomplished attorney, economist, and actor,[13] is also a proponent of the origins theory called intelligent design. The theory of intelligent design holds that certain features of the universe and of living things "are best explained by an intelligent cause, not an undirected process such as natural selection."[14] Those who believe in intelligent design do not necessarily believe in a young earth.

In 2008, Stern exposed some well-known scientists as they tried to explain their views on how life began. He also interviewed researchers who were dismissed from their positions because their discoveries and opinions didn't comply with acceptable conclusions.[15]

A more recent Creation documentary, *Is Genesis History?* took a positive approach showing how real science refutes long-standing thinking on evolution.[16] There were two primary ways this happened. The first is by actually looking at evidence on its own merit and evaluating it without the erroneous preconceptions that it must somehow prove that life has been developing here for hundreds of millions of years. The second is by using advanced new science. It puts new light on what older methods may have gotten wrong.

NEW SCIENCE TURNING BACK TO THE BIBLE

There are many examples of an unwillingness to accept Creation proof in science. For example, in 2005, molecular paleontologist Mary Schweitzer found elastic soft tissue in a T. rex bone. It was discovered in northeastern Montana at Hell's Creek. It was thought to be a 68-million-year-old fossil. Her boss became angry at her because her discovery would give credence to creationists. Since then, Schweitzer and others have found red blood cells, blood vessels, bone cells, and even hemoglobin and protein basic to animal tissue in fossils considered to be from 80- to 195-million years old.[17]

Scientist Mark Armitage was a microscopist on the electron and confocal microscope at the California State University at Northridge. In 2012, he published a paper about finding a 65-million-year-old Triceratops horn that was not fossilized. Neither was the soft tissue decayed. Within two weeks he lost his job but was later compensated. Armitage said, "soft tissue in dinosaur bones destroys deep time." They "cannot be old if they're full of soft tissue."[18]

Discoveries like this, using new research methods, are wiping out the multi-millions of years theories. As they do, they also wipe out the possibility that enough time has passed to be able to make the species-to-species changes that evolution requires.

We need to recognize in the Scriptures God already provided the key to understand. Animals did not evolve from one species to another but were made after their kind.

MAN WAS ALSO MADE, NOT EVOLVED

God does the same thing as we look at the creation of man. He said, "Let us make mankind in our image, in our likeness..."[19] We find this concept in different words four times in verses 26 and 27. The same literary tool that reveals the truth about animals procreating after their kind applies to humans. That is, with one very large exception. The beginning of our kind was special. We are created in God's image. Then, "When Adam had lived 130 years, he had a son in his own likeness, in his own image; and he named him Seth."[20] The Bible says children are made in the image of the parent, in the same way we are made in the image of God. Every succeeding generation is reproduced after the original, which is God.

If we are made in the image of God, and we are made in the image of our parents, we are different than the animals. It is as God says, "All these animals you see, all over the world, on the land, in the air, and in the waters, were made in this way, but you were not made like the animals! You are different than the animals. You were made in My image."

A HIGHER ORDER

We are a higher order in the scale of being than the animals, as we discussed in Day 5. But that is not all. Do you see something more in these verses? Do not forget this. Look at the text. He said, "Reign over the fish in the sea, the birds in the sky, and all the animals that scurry along the ground."[21]

Eight times in these three verses, God tells us that we are over all the other creatures. He didn't just say once, "You are over everything." He repeated it eight times to make sure we take notice.

I know of nowhere else in the Bible that God emphasizes an idea so many times in so few verses! It tells us the tremendous importance that God wants us to derive from this fact. This is important to God. There is nothing in the Creation narrative that God has emphasized the way He has emphasized this single point. We are over the other creatures in our biosphere.

Attention to this one concept is unprecedented in Scripture. He shows it in the order of Creation. He shows it in the attention that He gave to the way He made us. He says that we are in His likeness. Now, He comes right out and says, without mincing words, we are over all the other creatures. We have dominion over them.

This is the very question being asked today worldwide. People are being driven from certain habitats by governments on the basis of the presence of wildlife. We are to have dominion over the animals. This does not give permission for us to mistreat any creature. Still, the needs of humans take priority.

In 2014, a chimpanzee near Rochester, New York, named Tommy was represented by a group of attorneys to take its owner to court for mistreatment. An organization wanted Tommy to be granted personhood. The reason was to sue the chimpanzee's human owner in an attempt to gain the creature's own freedom.[22]

Of course, Tommy knew nothing about the case or his supposed benefactors. This seems novel to some. Had the case been decided in Tommy's favor, it would have set legal precedent for other animals to be used by militant attorneys to disintegrate the difference between people and the rest of the animal kingdom.

This case along with multitudes of human restrictions in favor of lesser creatures seeks to undermine the scale of being. It is another effort to remove God from His position at the top of the scale.

Had it been only a matter of mistreatment, laws already in effect could have been brought to bear to have Tommy freed. It shows a disturbing, strengthening trend to raise animals above people.

Where does it stop? It doesn't. The European Union began the process of granting the status of "electronic persons" to autonomous robots.[23] They speak of this as a protective measure to be sure these computerized bots can be controlled to protect the rights of people. Even with that assurance, the reported recommendations still use the terms "Frankenstein's monster, and the Greek myth of Pygmalion."[24] It seems that some are concerned that man's creation may do to us what God's creation has been doing to Him.

This may be more of a reality than many are willing to let on. In October 2017, "Saudi Arabia…granted citizenship to an artificially intelligent robot who once declared it hopes to 'destroy humans.'"[25] Some Saudi Muslim women complain that the bot has more rights than they do.

These examples of animal and electronic personhood exhibit the efforts of special interests to depreciate the value of what God values most, as well as the potential danger we are building for ourselves.

Chapter 9

Creation Day 6C

SPECIAL KIND OF CREATION

During Days 1-5 as God talked about what was going to be made, He said, "Let there be… and let the waters… and let the land."[1] The other things happened as He said they would happen. But on Day 6, when He was ready to create man, He said, "Let us make human beings in our image…"[2] At the word of God, nearly everything else was made as if by spontaneous generation. Not so for man.

The creation of each level of the scale of being became more complicated and more personal. The birds and the animals of the land seem to be on a higher rank on the scale than fish and water creatures. We are told "the Lord God had formed out of the ground all the wild animals and all the birds in the sky."[3] The fish were left out of this description.

God made man personally. Creation of man was unique showing the greater love of God for man. There is no indication that God spent more time making Adam than He did making the millions of fish, the millions of birds, and the millions of land animals. But as this list shows, there were untold millions of those other creatures. Each group of them was made in about the same time frame as one man and one woman. Another major difference between the land animals and birds, and man is the addition of God personally breathing life into him. "Then the Lord God formed the man from the dust of the ground. He breathed the breath of life into the man's nostrils, and the man became a living person."[4]

This is another indication of man's rank in the scale of being. God didn't simply speak man into being or impersonally create millions of individuals at that time and call them man. God bent down in the dust of the ground and formed the body of man with His own hands. When the Master Potter was satisfied, He bent down further and breathed life into the nostrils of man.

Man's life came directly from God. It wasn't by decree. It wasn't en masse like the birds and fish. It was one-on-one. It was time taken. God who controlled the universe took time to personally give life to one man. It wasn't a thoughtless act. It was intentional. While all life comes from God, our life and existence come with no intermediary, no space between. It was given by One who loved man and with the very highest aspiration for mankind—for you. God wants only the best for you.

God gave life to man. When sin entered the world, man essentially gave up life. He gave away the ability to live by turning away from God, the Life-Giver. God gave it back, not willing for us to be lost. Doing so cost Him immensely. "For this is how God loved the world: He gave his one and only Son, so that everyone who believes in him will not perish but have eternal life. God sent his Son into the world not to judge the world, but to save the world through him."[5] That is the reason Jesus came. It is so that you can be saved.

In the texts that describe God making man, there are additional important ideas we need to garner. First, God created man to be male and female. "God created man in His own image, in the image of God He created him, male and female He created them."[6]

The male, Adam, was created first. He spent some time alone, naming animals. Adam knew that he was alone. God had already prepared the plan to resolve this need. "Then the Lord God said, "It is not good for the man to be alone. I will make a helper who is just right for him."[7]

"So the Lord God caused the man to fall into a deep sleep. While the man slept, the Lord God took out one of the man's ribs and closed up the opening. Then the Lord God made a woman from the rib, and he brought her to the man.[8]

TWO WAYS TO CONSIDER RANK IN THE SCALE OF BEING

Let me remind you of some things we already covered. We have looked at the rank in the scale of being based on two things. One is "the measure of divine attention bestowed on [the] object..."[9] Mankind is the highest rank of all creation. Adam was made by God's special attention of kneeling in the dirt and forming him. Then, God breathed life into him. This makes the highest measure of divine attention having been bestowed on Adam.

Until Eve! God took the male on whose creation He lavished more attention than anything else. From this most precious creature, God took a part and fashioned another whole person—the woman, Eve. By our definition, this makes the woman to be the object of the highest measure of divine attention.

The second observation of rank is that God created each element and creature in an orderly fashion. He made things that would be needed by other objects which He would create later. For example, before He made fish, He made water. Both of these considerations seem to point to one huge question. Does this make woman to be a higher order on the rank of the scale of being than man? According to one of our rules for the order of creation, that which was needed by something else would be made first. Does the fact that man was made first mean that woman who would come later, would need man?

May I be honest about this idea? When you read the Bible and you find things that stick out and seem to be different than what everyone else is saying, you must say, "I've got to be careful with this. This could be heresy." I continued to study because I wanted to see if anybody else had ever said anything so heretical.

The Talmud is one of the two most important books of Scripture to the Jews. It says, "Eve, since she was created after Adam, is considered in Jewish thought to represent a higher form of life than Adam, since she was able to carry a fetus in her body."[10] Little more is said about this. A great deal of additional study is needed to hazard an opinion.

When Genesis talks about the animals, while they, too are distinguished as male and female; the Bible doesn't differentiate between them. Attention to this distinction made between male and female is only shown with mankind.

WHY DOES MALE AND FEMALE MATTER?

This is the only combination God recognizes to be in His image. He didn't give any other combination that was acceptable to Him. God gives each person the freedom to determine their own lives on this earth. But if we accept that male and male, or female and female can join together sexually, then the image of God becomes corrupted in our own minds and in the minds of others who may not grasp the significance of His image.

Among other things, the significance is that God made us in many ways like Himself. This is discussed in other chapters to different degrees. His instructions to Adam and Eve may bear a clue to the reason God favored humanity so highly.

When we are told that we were created in His image, being male and female was singled out as a part of that. Then, "God blessed them and said, 'Be fruitful and multiply. Fill the earth and govern it. Reign over the fish in the sea, the birds in the sky, and all the animals that scurry along the ground.'"[11] He trusted us to carry on His work of filling the earth with people who He would love and who would love Him. This is a huge responsibility. It is one that continues today and was reiterated by Jesus.[12] It is the kind of heritage that is left to sons and daughters who are in His image.

God wanted His earth to be taken care of the way He would do it. That would guarantee the continuity of its beauty and utility. He built in us the characteristics of Himself that would be able to do this work. This also applies to everything else He asked of people later.

Part of God's image in every person is a reminder to themselves that they are being called by Him. By our lives and actions, we are to show others who God is, and lead them to Him. We are the only picture of God that some will see. It needs to be a representative picture to give others the right perspective.

Let me make a disclaimer here. There is not a person living, nor dead, whom God does not love more than we have the ability to comprehend. Every person who breathes struggles with one issue or another—most often multiple issues.

The fact that we have struggles says nothing about God's love for us or about our potential for salvation. Rather, it says that we are human. Human is good. To be human is to be at the top of the Scale of Being. However, because of the sin of Adam and Eve, we have all fallen to a dreadful condition. "All have sinned and fall short of the glory of God."[13] The condition of mankind is so bad that "no one is righteous—not even one."[14] This puts us all in the same boat. God has shown us the problem of sin, and the solution. "The wages of sin is death, but the gift of God is eternal life in Christ Jesus our Lord."[15]

PARADOX OF A LOVING GOD

Consider this. Eve didn't sin because she ate fruit from a tree. Eve's action merely verified the sin she had already committed in her heart. Eve made a conscious decision to disregard what God had said to her and surrender her will to the contradictory voice of another. This is the foundation of the sin of every one of us. The misplaced surrender to a voice contrary to God becomes sin long before we commit the deeds that verify it.

Force or punishment cannot dissuade conscious decisions. Often choices we make are promoted within us by unconscious stimuli. These have been placed in our minds by our actions and desires, and by the evil one himself to intentionally hurt us. But God has a solution. He "showed his great love for us by sending Christ to die for us while we were still sinners."[16] This way, He "freely makes us right in his sight" and "freed us from the penalty for our sins."[17] This gift is available to every person who is willing to surrender to loving God.

We cannot change the Word of God nor the message of it. Being created as male or female is a very specific part of what it means to be in the image of God. Public sentiment in the United States and much of the world today is dramatically moving toward acceptance of lifestyles that are contrary to God's design for His people.

OTHER SEXUAL LIFESTYLES

For the sake of simplicity, I will use the term "GSM" to represent the ever-lengthening list of non-heterosexual and non-cisgender people.

GSM stands for gender and sexual minorities.[18] Gallup research in January 2017 reported that 4.1 percent of American adults identified as gay, lesbian, or bisexual.[19] It often seems much larger because of the popularization by media and the strong political status of the community. All are precious in God's sight.

I have no intent to belittle GSMs, nor to seek to control them or change them, unless they wish to be changed. But let me be clear. Every person has been given the freedom by God to make a decision to do anything or live any way you wish. I am not challenging that whatsoever. Each of us is free, but we are seldom free from the consequences of our choices.

I recognize that this passage of Scripture is very restrictive as to what kind of sexual relationship represents the image or likeness of God. It also means that people in heterosexual relationships or activities other than lifelong, monogamous, committed relationships are not living according to the image or likeness of God. (Remarriage after the death of a spouse is a biblical exception.) For any of us to base our identity on our sexuality is to deny God's image in us. If our sexual identity differs from the way He says He made us, then we are denying God except where we may seek to manipulate God's image to suit ourselves. This is very serious.

It must be noted that sexual lifestyle is not the only characterization of God's image or likeness. It is just one that was singled out in the Creation narrative. That may be because it would be a major issue at the end of time. It is not a matter of personal preference or what people think or how they feel. It is a matter of what God has said in Scripture. Only a male and a female uniting in marriage is in the image of God.

The image of God is important because it must be recreated in us for us to gain entry to Heaven. Our own efforts to recreate God's image are idolatry. We cannot do it. Only God can renew His image in us. There are 1,189 chapters in the Bible. Two of them describe God's Creation to give us His image and character for eternity. One chapter speaks of the fall of mankind into sin. The other 1,186 tell of God's efforts to renew His image in each of us so we can spend eternity with Him.

MARRIAGE DISPLAYS THE IMAGE OF GOD

In Scripture, God used marriage to show the relationship between Himself and His people. The Old Testament book of Hosea illustrated the imagery of God being a husband to Israel. It is clear in the New Testament. "Husbands, this means love your wives, just as Christ loved the church. He gave up his life for her to make her holy and clean, washed by the cleansing of God's word."[20]

Husbands need to love their wives so much that they are willing to die for them, if that is what it takes for them to be able to be found in the kingdom. But it also says, "He did this to present her to himself as a glorious church without a spot or wrinkle or any other blemish. Instead, she will be holy and without fault."[21] The metaphor shows Jesus doing what is necessary to save us, His bride. A human husband cannot save his wife, but his relationship with her should lead her to be ready for Heaven.

Marriage was pure and holy. God made it that way. Marriage represents God. It was intended to be a joining of man and woman and God. It represents His Trinity, the Father, the Son, and the Holy Spirit. In human marriage, there is the husband and wife and God. It is a trinity of humanity with God. His plan was that every time we see a married couple they remind us of God.

In marriage, we represent the character of God. The Bible places a high expectation on marriage. It is the reason He asks us to also place a high expectation on marriage. When we enter into marriage, God asks us to be sure we are doing the right thing. This high expectation of God for marriage may also play into why some are dedicated to destroying biblical marriage.

The Talmud has compelling evidence. Here is the way God sees male and female marriage. "The Talmud says that the Hebrew words for man, ish, and woman, ishah, are identical, except for the letter yod in *ish*, and the letter hay in *ishah*. The two letters, yod and hay, together make up a name of God."[22]

The joining of man and woman together forms the image of God in the language of the people He chose. This is no accident. If we join man with man, or woman with woman, it makes a false image of God, which is idolatry.

In Day 5 we learned of a warning in Revelation 14:7 that during the last days, men will challenge the very concept of God's existence by doubting Creation. As we study the creation of man, without doubt, many of today's attacks on marriage are specifically to destroy the concept of the existence of God. It is possible that some who participate in those attacks may not really understand the reason why they are doing it. But the devil is leading them. Their end goal is to prove that God does not exist. They may settle for showing that God and His image are not important.

Real Christians accept all people. It is because all were made by God to be in His image. Sin in the world caused all of us to fall out of being in His image. It is God's plan to return every one of us to be in His image, no matter why we become separated from Him. His love for every person is the same as for every other person. Here then we see the danger of the true intent of the downward evolution of man in our day.

Some who believe that they are gay or lesbian, or another of the GSM community remain celibate. They choose not to participate in the activities of the lifestyle. This is very difficult to do. It is similar to the person who has strong heterosexual drives but is not married. That person also is to remain celibate and refrain from involvement in impure lifestyle activities. Both may have pressures from their own desires as well as from other people. They can live without sinning by not acting on their desires and feelings. A person who feels gay is not sinning if they do not act on that feeling or focus on it mentally.

This is not an attempt to attack persons who find themselves to be gay or a person of any other identity from the GSM community. God loves each one the same as He loves those who are straight.

OTHER FORMS OF IDOLATRY

There are other sexual forms of image-of-God idolatry. These things are also against the image of God or the character of God. They include sex with someone other than the one you are married to in all its variations—premarital sex, post-marital sex, multiple partners, self-gratification, bestiality, incest, and many more.

God does not change. Our role in the image of God in our sexuality will not change. We will have a sexual role with only one person in our lives, except in the case of death of a spouse.

A person may not have understood the significance of our God-likeness. But if we decide that it doesn't matter that we promised God to stay married to the same person until death, is it possible we are saying that being in the form of God is not important to us. Our feelings are more important.

Often, teens and young adults decide to become involved with sex before and outside of marriage. This says, "Being in the image of God is not important to us. Our fun and our reputation with our friends is more important." Those who refuse to be made in the image of God will not be ready for heaven.

Whatever sexual departures we may be involved with, God still loves us." We have all made idolatry of our lives, and we can come to Him and ask Him for forgiveness. No matter what our sins have been. God is forgiving. It is a part of who He is as God. He does not simply forgive, but He is forgiving. He is forgiveness. "If we confess our sins He is faithful and just to forgive us our sins and to cleanse us from all unrighteousness"[23] He makes new creatures of us. Jesus tells us, "Neither do I condemn you… Go now and leave your life of sin."[24]

BLESSING AND COMMISSIONING

God blessed and commissioned male and female. This is part of the reason why other sexual combinations of people are contrary to God's will. There aren't any others who can carry out the primary commission. "God blessed them and said to them, 'Be fruitful and increase in number; fill the earth and subdue it. Rule over the fish in the sea and the birds in the sky and over every living creature that moves on the ground.'"[25] This inability to procreate naturally, however, does not keep anyone from being loved by God.

We have been given the privilege to reproduce at a level higher than other creatures. This includes being given the task of filling the earth with people who loved God and who would be loved by Him. The commission also demands that mankind take care of their families. They are to train their children to know, to love, and to serve God.[26] It is parents, and not any government, who have the right and God-given responsibility to determine how children are raised and what their standards will be.

THE CREATION OF BIRTH

We have thus far presumed something which may not be the case—that birth was a normal process at the time of Creation. For animals and man to reproduce after their kind, and for man to be reproduced in the image of God, there had to be a process for it to be accomplished.

Consider that in heaven there had never been a birth. Quite likely, it is the same on other inhabited planets. There is no need for birth in any of those places because sin doesn't exist there. Only Earth defected from the rule of God by sinning. Any other place in the universe where there is life, it was all created. Since there is no sin, there is no death. So, there was no need to make provision for sin's consequences. No need to replace those who died to continue the species. This is true of plants and animals—every living thing.

We already discussed seeds in Day 3. They were placed in plants for those forms of life to be continued until the rescue mission, i.e., the plan of salvation, was fulfilled. Seeds also pointed forward to the birth of Jesus to return us to the image of God, and to immortality. That was only needed on Earth.

Each succeeding higher level of the scale of being reproduced at a higher level of complication and involvement than the one before it. There may be limited exceptions at every level. Single cell life forms simply divide. This does not suggest that such cell division is a simple process, but that it is done with no planning or discretion. Plants dropped their seeds which have been pollinated with the help other creatures or by wind. This is done without any effort on the part of the plants. There is no relationship between the seed producer and the seedling.

Fish generally lay their eggs in the water, and they are fertilized in the water. This is typically accomplished without any real communication between the male and female. Certainly, there is no relationship developed. These eggs are lain by the hundreds and are rarely cared for by a parent.

Reptiles have a seasonal mating period. All mating is done by instinct and not by choice.

Birds lay fewer eggs, but usually more than one. A mating period often attracts a suitable mate. When the right one is found, fertilization is performed. One parent typically guards and feeds the offspring, but leaves them when they are ready to fly from the nest.

Insects, fish, birds, reptiles, and land animals mate by instinct. The lower they are on the scale of being, the fewer mate for life.

Humans are intended to be mated for life. A strong interpersonal relationship is intended to benefit the family, including the mate and children, even to several generations. We mate, not by instinct, but according to the desire built by the relationship. Biological processes and the personal attraction to another are part of human mating. This is intended to be controlled by the couple and limited to one mate.

GREATER REASON FOR BIRTH

Any such plan for people had to be as readily available to them as was the birth process. That means salvation is available to every person who was born. God needed a method to insert Himself into the daily life of man.

Birth provided the mechanism to bring salvation to people in a way that it might appeal to every individual in a personal way. The plan had several requirements. The Savior had to be born as we are born but be able to show that He was special. Matthew exclaims, "'Look! The virgin will conceive a child! She will give birth to a son, and they will call him Immanuel, which means God is with us.'"[27]

He had to be "born of a woman, subject to the law."[28] He had to be "tempted in every way, just as we are—yet...not sin."[29] His birth, life, and death would show God's great love for us. "For this is how God loved the world: He gave his one and only Son, so that everyone who believes in him will not perish but have eternal life."[30] In this, it also allowed for the free will of each person not to accept Him if they so choose.

Equally important, this would have to be evident through the rest of Creation, even if a person never heard the name of Jesus. "For ever since the world was created, people have seen the earth and sky. Through everything God made, they can clearly see his invisible qualities—his eternal power and divine nature. So they have no excuse for not knowing God."[31]

MOVEMENTS CHALLENGE MAN'S RANK AND GOD'S IMAGE

Today's popular movements seek to turn the rank upside down and, in the eyes of man, change the value of what God created. Why? In Creation, man is not truly the top rank of the scale of being. God is at the top. This is the reason Revelation instructs us to worship God as Creator.[32] Proper worship keeps the scale of being in the correct order.

This is God's order in the rank of the scale of being.

God

Man

Animals

Fish, Birds

Sun, Moon, Stars

Plants

Air, Atmosphere

Water

Dirt, Elements

Evolution has long taught that God had nothing to do with the development of the universe or of man. It is clear to see that as a blatant attack. This puts God at the bottom instead of at the top—or removes Him altogether.

The dramatic trend to normalize homosexuality is a clear attack on the family, on marriage. It directly challenges the image of God, and thus His position as Creator. We must be ready to minister to and love those who self-identify as GSM. Straight people are not better people than GSM's. But we are not to allow that ministry to support the destruction of the image of God in His people.

Environmental and animal rights movements strive to destroy the blessing and commission God gave to man and to woman. There is evidence that portions of the environmental movement abuse science to manipulate data to prove their points.[33] Recently some have seen that the purpose of abortion is not generally to protect the life and health of women. This flagrant disregard for the unborn is to show that God-given life has no value other than monetary or for social engineering.[34] Life is believed by many not to be from God at all.

Add to these the many other factors that are working to remove God from our world—fascism, socialism, communism, education in some subject areas, news media, entertainment, so-called science, feminism, religious extremism, politics, and religious ecumenism for the purpose of unity.

No wonder Jesus told us to be careful about the end-time mindset and instructed us to remember to continue to worship Him as Creator.

God made man because of His love for you. Long ago the Lord said to Israel: "I have loved you, my people, with an everlasting love. With unfailing love I have drawn you to myself."[35]

Since God planned you, made you, and loves you, He will not abandon you or turn away from you. You can turn from Him, but He will be right there, longing for you to turn back. He has said that.

You are exactly what God wanted, except for the sin in your life. He will build you back to His image if you will permit it. He is ready to begin right now. Find joy in the way God is leading in your life, but be sure it is God who is leading and not some other force or power. God knew what life on Earth would be like, just before His return. Only He can renew us in His image and prepare us to live forever.

Do you want to be ready to meet Jesus when He returns? I'd like to ask you this question seriously. It is the same question that was asked by the Apostle Paul. What will you do with this Jesus Who has made you?

Chapter 10

Creation, Day 7A

PORTAL TO ETERNITY

Dateline Phnom Penh, Cambodia, 1973. My work as director of the English Language Center in Phnom Penh and supplying needed resources to refugees brought me into contact with government officials. My relationship with them often provided invitations to visit various sites throughout the country.

Since my arrival in Phnom Penh, I had heard rumors about the existence of twin pagodas—one silver and one gold. The possibility of such sites fascinated me. I couldn't find out where they were. I asked several local people whom I knew. Some told me that they had heard the rumors too but believed that they were only stories to entice the tourists. These famous buildings were considered sacred sites and were said to be located in the capital city. I never dreamed I would have the opportunity to find, much less to visit either of these pagodas personally.

I don't remember how the invitation came. Perhaps I finally asked the right person and he had connections. My unofficial invitation to visit the Silver Pagoda was delivered.

This pagoda was a holy place to the Buddhists. They set the rules for visitors to enter. Upon my arrival at the entrance to the

Silver Pagoda, I knew a protocol had to be followed. They set the rules for being there.

Every item in your pockets had to be removed and placed in a small bowl. You couldn't even have a comb in your pocket. Attendants kept these items secure until your departure.

Everyone had to take their shoes off. Upon entering the Silver Pagoda, we walked on a floor of silver.

In the front of the prayer hall, many statues of the Buddha were placed on varying levels of platforms like silent sentinels. Guests and worshippers sat on the floor. Pockets emptied of their contents prevented visitors from scratching anything—especially the floor. No one carried a tool that allowed them to etch out the silver. Such behavior would not be acceptable in the temple. In a time when the Khmer Rouge were surrounding the city of Phnom Penh, everybody was looking for whatever form of currency they could find. This place would have been a tempting hunting ground.

When I entered the Pagoda, I took my shoes off. Was I going in to worship the Buddha? No. But because I had been invited to come to this extraordinary shrine, I demonstrated respect for the beliefs of my hosts. I removed everything out of my pockets. I asked the guard for permission to carry a small camera with me. Of course, that was before cell phone cameras. Imagine my surprise when I received permission to do so.

I sat quietly on the floor, making sure the bottoms of my feet pointed away from the statues. This was out of respect for the people who believed in Buddha. There was no god there to worship. There were only statues made of wood, ceramics, stone, silver, and gold.

Did you notice the care these people used in the way their god is worshipped? If the temple had been a Christian church, God would be glorified by that special attention!

God made a temple. Yes, we may worship each week in a church building. But we recognize that the church is the people and not the building. The building where we may meet, not just one day a week, but any other time as well, is not holy. The furniture is not holy. The carpet is not holy. The place itself is not holy. The God we worship is holy. His presence is what makes our place of meeting a special place. The time we spend worshipping with Him can be holy. A physical temple can be easily ignored or avoided—not so with time.

HOLY TIME

Time was established on Day 1, "the evening and the morning were the first day." Genesis speaks of time matter-of-factly. It simply was. In the rank in the scale of being, there was nothing created on Day 7. However, there was an even higher status given to a specific portion of something that had already been made.

At the end of Day 7, after every thing had been created, a specific one of the "evening-and-mornings" was made holy. The hours of Day 7 were placed on a level all their own. Not all time was made holy. Just that specific seventh day. It became holy because the One who sanctified it is holy and chose to make it holy.

The separation of days and nights was finished; the firmament was finished. The gathering of the waters from the dry land, the grass and herbs, the fruit trees, the sun, moon, and stars were finished. The birds above, the sea creatures and everything that moves in the waters; the cattle and the creeping things, and finally man and woman were finished. It was all good. There was nothing else that was needed. Except…the Sabbath, which is one of our measures of time. Time is among our greatest gifts. Often, we don't seem to have enough of it. On the other hand, sometimes it seems to drag on too long.

Every second of time that ticks away promises that there will be an end to suffering and sin. Therefore, all time, and particularly "the Sabbath was made for man."[1] It is for our benefit so that we will not be forced to live forever in the condition of sin and the corruption it causes. The seventh day was just another day that naturally followed the day before it. Being the seventh day was not what made it holy. "God blessed the seventh day and made it holy, because on it he rested from all the work of creating that he had done."[2] "He made it holy after He had rested in it."[3]

God focuses our attention on the Sabbath as on no other day of Creation, so we can begin to understand what a great glory it is for us to have time. We noticed at the beginning of Day 1 that God created time to limit the reign of sin in our universe. It needs to be reiterated here.

Lucifer, the covering cherub, rebelled against God and His authority. How would God put an end to this rebellion? He might have simply destroyed Lucifer. But short of destroying Lucifer and the other angels who rebelled, what else could have been done? How can you limit the effects of a rebellion in eternity?

Free will allows for rebellion. God permitted rebellion—even provided for it. But He wanted to put a time limit on it! Sin had to be curtailed from extending throughout the universe and existing for eternity. It had to be put into a box to contain it. That box is time. But until the beginning of the Creation of our world, time didn't exist. That is why God created it.

TIME—AND HOLY TIME

The Sabbath is the one day of Creation that points specifically to time. We need a regular and frequent reminder to put God first. You and I become so entertained by the nervous jittering of our daily existence, we forget salvation is fast approaching. We need a consistent and frequent reminder to put God first. Only a regular, special marker in time can help us not become too busy with our daily activities. Otherwise, we focus on the things that rush in on us to the place where one day after another becomes the same. So much needs to be done. Many never take the time we need to spend with God.

"So the creation of the heavens and the earth and everything in them was completed. On the seventh day God had finished his work of creation, so he rested from all his work. And God blessed the seventh day and declared it holy, because it was the day when he rested from all his work of creation."[4]

TIME TO WORSHIP GOD

God made this period of time holy, a place in time where we are to worship Him. He made it a temple. We are invited to worship the God of that temple.

God created this holy time on Day 7 to be a time when mankind can recuperate from the burdens of the previous six days. As you lay aside those things, your mind and your body can rest and recuperate.

I could have declined the invitation to go to the Silver Pagoda. Many people you invite to come to your place of worship decline the invitation. But no one can refuse to enter the temple God made. We cannot choose to avoid the holy Sabbath Temple. God made it so that everyone must go through it to continue to live another week. Every person who has ever lived seven days has worshipped in this temple.

The one you choose to worship is not determined by being in a place, or even by being in a time. Worship is what your heart and your mind are doing during that worship experience. Although I entered the Silver Pagoda with respect to those who invited me there, I did not worship Buddha. I was impressed with the lengths people had taken to glorify a dead person but remembered to direct my thoughts to the living Creator. In the temple of time each seventh day, everyone enters. Everyone worships. Some worship God.

If everyone worships, and only some worship God, who are the others worshipping? They worship the devil. There are only two choices. A physical temple can easily be ignored. You can choose to go in or not. But everyone who goes into God's time memorial of Creation either worships the Creator, or they desecrate the temple and worship the devil. There's not a third option.

No one who lives through a week avoids the passage through God's holy day. A person may come eagerly, fearfully, or even ignorantly. But we will meet God in that time. He will be present with us in His holy temple.

WHAT GOD ASKS OF US

What are God's expectations of us then, when we are in His memorial temple? "If you turn away your foot from the Sabbath, from doing your pleasure on my holy day…"[5] Now here's the difference between the way some people look at it. Some people would read the text as if it says, "I will give God an hour in the morning on Sabbath. I'll come during that hour, sit quietly, and think about what I'm going to do for the rest of the day. It may be the football game or the beach where I'll take my family. After all, it was made for me."

But God says, "… turn away your foot from the Sabbath…," (not part of the Sabbath, but all the Sabbath) "from doing your pleasure." What is your pleasure? My pleasure is the non-God things I think about while in His holy time temple.

But remember, you are not primarily in a place when you are worshipping God. You are in a time. You have entered a time of very discernible beginnings and endings—the setting of the sun on the sixth day, and the setting of the sun on the seventh day.[6] God is not asking us not to do our own pleasure on His holy hour, but His holy day. If during the Sabbath we are spending the time saying, "I wish this time would end so I can get on to what I want to do," we are still not worshipping God.

Jesus even spoke of this in a conversation with a Samaritan woman. He told her, "The time is coming—indeed it's here now—when true worshipers will worship the Father in spirit and in truth. The Father is looking for those who will worship him that way. For God is Spirit, so those who worship him must worship in spirit and in truth."[7]

PHYSICAL AND SPIRITUAL RENEWAL

The Sabbath is not a physical thing, but a spiritual element. We are certainly physical, but we are also spiritual. We can see the physical side of ourselves. It has deteriorated due to the sin of the world. Our physical and spiritual side were both created in the image of God. There may or may not still be some aspect of our physical selves that remains in His image. We will know when we see Him.

Our spiritual side has most certainly deteriorated dramatically. Each person who depends on following the spiritual lead of their culture, or of their family is in danger of following the spiritual mutation of generations before you. It is by being personally renewed by the Holy Spirit and by the Word that we have a fresh spiritual connection with God.

It is during the Sabbath that God especially seeks to renew our spiritual selves in His image by spending the whole day with us once every seven days. During this time, He will build a relationship with us in His temple. He can renew His image in us because "if we confess our sins to him, he is faithful and just to forgive us our sins and to cleanse us from all wickedness."[8] Naturally, God can do this for us any day. But it is more likely to happen for us when we are concentrating our attention on Him for an entire twenty-four-hour period. He is available for us personally during that time in His temple.

Our bodies will also be renewed in God's image. When Jesus returns to get His people who have allowed Him to renew them spiritually, "our dying bodies [will] be transformed into bodies that will never die; our mortal bodies [will] be transformed into immortal bodies."[9] They will not only be immortal bodies, but "The Lord Jesus Christ, who, by the power that enables him to bring everything under his control, will transform our lowly bodies so that they will be like his glorious body."[10] The renewal of the Image of God in man will be complete. It is all His work. As He made man at Creation, He is renewing us even now. He does the work in us that we cannot do in ourselves.

The rewards of allowing Him to renew us are great. "Then the Lord will be your delight. I will give you great honor and satisfy you with the inheritance I promised to your ancestor, Jacob. I, the Lord, have spoken!"[11]

God is very clear about what He wants from the Sabbath. Why is He so clear? He wants us to focus on Him. In the Creation week, God made everything, and then on Sabbath He came to be with Adam and Eve. He modeled the purpose of the Sabbath. He designed it for us to look forward to the entire week. Even though we experience problems, trials, and troubles we can come to this temple of time and know that we are meeting God.

READY TO MEET THE KING

I've observed the way people prepare to meet presidents. I've seen the anxiety of preparing to go into the presence of kings and queens. In Thailand, I was once summoned to meet the personal representative of the queen. In the Congo, I met with the First Lady to televise the work we were doing in their country. For both events, I didn't go casually. When we come to meet God Himself during these hours of the Sabbath, do we come on His level? Do we come perfect and holy? No. We come as sinners.

One of the amazing things we can recognize is God calls us to come and visit with Him personally because He knows we are sinners. Why would He want to meet with us? The Sabbath is the time God instituted for us to be able to spend with Him and build a relationship. That is what you do when you spend time with other people. You build your relationship with them. If you neglect to spend time together, the relationship slowly dies. Crooner Dean Martin is credited for saying, "Absence makes the heart grow fonder (for somebody else)."[12] God wanted our relationship with Him to live and grow. Remember, "the Sabbath was made for man."[13] It was made for us to be with Him. It is not special because we are there, but because He is there with us!

When we become intentional with God, and our relationship with Him grows, and He will have the opportunity to re-create His image in us. That is not the only reason He wants to be with us. He also wants it simply because He loves us! He enjoys our company.

THE VALUE OF ONE DAY

God loves us so much He intentionally placed an entire 24-hour period for us to spend in His presence. We noticed during the days of Creation how much can be accomplished in 24 hours. The earth can be cast into space in just 24 hours. The waters can be divided from the waters and the firmament can be built in 24 hours. The entire earth can be beautifully landscaped in just a single day. The sun and the moon can be created in 24 hours. Birds and fish can be scattered all over the world in 24 hours. Animals can be placed on every continent and man can be created—all in one day.

What can we do with 24 hours of time? We can worship the One who created everything. He gives us the gift of the Sabbath to say, "Watch what I can do for you in just 24 hours!"

God calls us. He invites us. In fact, He requires us to go through His temple of time, no matter what. He invites us to be able to spend 24 hours enjoying His companionship while being re-created in His image. If man had always chosen to spend Sabbath with God, instead of against God, then the world wouldn't be the way it is today.

We can choose to spend the Sabbath with God, making decisions every moment of the day based on what will bring us closer to Him. Or we can spend the Sabbath thinking, "How close can I get to the devil without actually worshipping him on the Sabbath?"

Learn the models of Creation. This is the reason why we've spent so much time looking at each day of creation. During every day you live, and in everything you see, God is reminding you to be ready to worship Him, particularly on the Sabbath.

Remember the fact that the sun comes back every morning? It is a promise that Jesus is coming. Remember the fact that there is water in the oceans? This points to Jesus, the Water of Life. Remember the seeds in the plants? Every seed is a promise that Jesus would come as the fulfillment of two promises. The first is that a Savior would defeat the evil one for us. God said, "And I will cause hostility between you (Satan) and the woman, and between your offspring and her offspring. He will strike your head, and you will strike his heel."[14] The second promise was made by Jesus, Himself. "When everything is ready, I will come and get you, so that you will always be with me where I am."[15]

Remember birth? God made and gave us the privilege to be a part of it. Every child we see is the promise of a Savior. Remember the male and female relationships? God led Adam and Eve to come together in a certain way and every married couple we see should reveal God to us. All of creation is made to continually remind you God loves you. He calls you to prepare to come to Him on the Sabbath day and be renewed in Him.

It is an amazing thing. The Sabbath was made for man. The Sabbath was made for you. There is nothing else that man devised since Creation that comes anywhere near taking the place of spending 24 hours with God as He renews His image in us.

Chapter 11

Creation, Day 7B

CREATING THE PORTAL

What it must have been like as the sun rose on the morning of the seventh day! Crisp, clear air greeted Adam and Eve. They drank cool, refreshing water. Brilliant hues of red, yellow, green, purple, and orange in the plants and trees attracted the eye. The scent of the flowers tantalized the olfactory nerves. The sounds of nature abounded. No one had any fears. The great animals roamed with the tiny ones, without suspicion or concern. The man and the woman were comfortable with each other yet tantalized for having received the abundant gifts of all creation, especially the gift of each other and for having known each other.

Could there be any greater day in all the history of the world? Words of poets and prophets fail to reveal the grandeur and the intimacy of this new world. It was made during the past six days, just for these two people. And, it was in itself, a promise to Adam and Eve of what God had in mind for every other person they would bring into it.

Synonyms of "special" do not begin to relay its reality. Words like exceptional, distinctive, outstanding, singular, notable, and remarkable don't touch it.

INGREDIENTS TO MAKE A HOLY DAY

Then, God put His final touch on Creation. Notice again the description of Day 7. "So the creation of the heavens and the earth and everything in them was completed. On the seventh day God had finished his work of creation, so he rested from all his work. And God blessed the seventh day and declared it holy, because it was the day when he rested from all his work of creation."[1] There are four things the passage tells us God did.

He rested on the seventh day.

He blessed the seventh day.

He sanctified the seventh day.

He made the seventh day a memorial.

Consider what each of these means.

1. God Rested on the Seventh Day.

God ended His work which He had done, including "the heavens and the earth and all the host of them."[2] This gives the impression that Earth was the final act of Creation in the universe. With the creation of this earth, all the heavens were finished. God wasn't tired. He didn't need to rest. But the Sabbath was made for man![3]

Man didn't need physical rest that first Sabbath. He had been alive only a few hours! But as usual, God created things before they would be needed. He knew that when man sinned, his labor would become difficult instead of pleasant and invigorating. He would welcome the day of rest after six days of work.

There is no indication of any global creative activity from Day 7 until the time still in our future when God will recreate the world. "Then I saw a new heaven and a new earth, for the old heaven and the old earth had disappeared. And the sea was also gone."[4]

What does God mean by this? He is talking about the time in the future when all these hours, minutes, and seconds have passed. God will wipe away all the sin and sadness, all the tears and pain. Everything that might remind us of the sin of the past will be gone, except the scars in the hands of Jesus. He will recreate the earth and the heavens so that we have a new place with no sin, no sorrow.[5] He is looking forward to that day. We are too.

The seventh day rest points us to a future moment. The only creative work to be done during this interim between the first Creation and the recreation is the transformation of sinners into new creatures. That is something God finds happiness in doing for each of us who will allow Him to. It is when He renews us in His own image. He wants to do this for every one of us.

Humans need to "lay aside their own interests and pursuits for one full day of the seven. By doing so, we can more fully contemplate the works of God and meditate on His power and goodness. We need it to remind us more vividly of God and to awaken gratitude. All that we enjoy and possess comes from the unselfish hand of the Creator."[6]

2. God Blessed the Seventh Day.

What does it mean to bless the Sabbath? What other things did God bless? Look at creation itself. Everything that God made during the six days, He called good.

There are only two times God gave a blessing during Creation week. He blessed the water creatures and the winged birds,[7] telling them to be fruitful and multiply. This is the same blessing He gave to the male and female, the crowning acts of His creation.

To the blessing of mankind, God added a commission. There was something He wanted us to do. In other words, God gave us a job. "God blessed them and said to them, 'Be fruitful and increase in number; fill the earth and subdue it. Rule over the fish in the sea and the birds in the sky and over every living creature that moves on the ground.'"[8] Mankind was the one thing God both blessed and commissioned. We alone are so special to God that He sacrificed His only Son to save us.

In the Christian era, God issued a portion of this same commission to His people in slightly different words. You see, His purpose was to fill the earth with people whom He could shower His love on, and who would love Him in return. That was the purpose of the commission He gave to Adam and Eve.

After His resurrection, Jesus gave His disciples their commission, which is also the charge He gives us. He used different words "Therefore, go and make disciples of all the nations, baptizing them in the name of the Father and the Son and the Holy Spirit. Teach these new disciples to obey all the commands I have given you. And be sure of this: I am with you always, even to the end of the age."[9]

The commission that God gave to everyone who accepts Him is, "Fill the earth with people upon whom I can shower my love and who will love Me in return." It is the same commission, both at Creation and at Christ's ascension. It is the invitation in the beginning for us to worship God as Creator, and the reminder by Jesus to do the same.

3. God Made the Seventh Day Holy.

"God … sanctified the seventh day…"[10] What does "sanctify" mean? It means to "make as dedicated to God; either in becoming more distinct, devoted, or morally pure." It is to cause something to change or to make something change.[11] God blessed the animals. He blessed and commissioned man and woman. But God blessed and made holy the Sabbath day. He changed the day from being just like any other day, to being holy.

Why didn't God make man holy? Because He didn't need to. Man was created in the image of God. God is holy, and His image is holy. God created man. Humans didn't have to be made holy until sin ripped holiness from them. Even today, the Christian who chooses to turn away from God and His holiness decline His justification and sanctification. Adam and Eve were created in God's holy image but chose to sin. They had to be covered by the coats of sacrificed animals.[12] This represents our need to be covered by the robe of the sacrificed Christ.[13]

If God did not make the Sabbath holy, it would have been like any other day. None of the other days were holy. The Sabbath was not holy until God called it holy.

Once something is made holy, the only way it can become not holy is for it to choose to be not holy. Man chose to be not holy by turning against the will of God. A day cannot choose to be not holy. It can't change itself. It cannot be changed by man.

The Sabbath was made for man. It was for man's benefit, but it is not under man's control. God made the Sabbath holy after He gave man dominion over all the created things. We do not have dominion over the Sabbath. It was made for us. God knew there would be attempts to change it. He didn't give man permission or authority or power to change it. To go against what God has made holy will only result in us going against God.

After sin, God had to do for man what He had done for the Sabbath day. Those who surrender to him, by grace that comes through the death of Jesus, God declares to be holy, the same way He did with the Sabbath. Once we have sinned, there is no other way for man to become holy again, other than God's declaration. We can't gain holiness by resisting sin.

Today the declaration of holiness for man is temporary. When we see our need of God, we confess our sins to Him and ask to be forgiven. "If we confess our sins, He is faithful and just to forgive us our sins and to cleanse us from all unrighteousness."[14]

The Bible calls this being justified. This is what happens when God says, "You have sin in your past, maybe even in your life. But you have surrendered to Me and I have called you holy. I have justified you because of the death of Jesus on the cross." God has the authority to do that. He paid the price for your sin and mine.

As we draw closer and closer to Jesus in our daily walk we surrender more and more to Him. The closer we come to Him the more we want to be like Him. As we behold Him, we become changed. As we spend time with Jesus, we surrender everything to Him. As each unholy desire is surrendered, God takes it away, each little portion at a time. By His work of grace, we become sanctified.

Sanctified is the word God used for the Sabbath! He sanctified the Sabbath, made it holy. In the same way we can't break the sanctity of the Sabbath, no one can break the sanctity of our sanctification. No force on earth can do that except our own stubborn choice.

God gives this wonderful promise. "I am convinced that nothing can ever separate us from God's love. Neither death nor life, neither angels nor demons, neither our fears for today nor our worries about tomorrow—not even the powers of hell can separate us from God's love. No power in the sky above or in the earth below—indeed, nothing in all creation will ever be able to separate us from the love of God that is revealed in Christ Jesus our Lord."[15]

Nothing can bring separation between you and God once He has brought you to Him and you have surrendered to Him. Now that is an important point because we must surrender to Him. If we come to the point in our lives where we decide we no longer want to be surrendered to Him, then it was nothing on earth that forced us away from God. We choose in the same way Adam and Eve chose in the Garden. They were holy and became lost. If it could happen to them, it can happen to us.

We have the power to do that. Why? Because God is not going to bully us. He will never compel anyone to continue to follow Him. That is against His character. That is against His love for us. He wants us to follow Him because we love Him. He wants us to choose heaven because we have made the decision that we choose to be there.

Although we can choose God, we cannot make ourselves holy. Even as God made a day holy, He can make you holy if you are surrendered to Him.

Some people believe that they are too bad for God to change or to accept them. The One who called the world into existence by His voice can save anyone who is willing to be saved. No one is so far gone that they cannot be redeemed if they will surrender to God. He can and will renew His image in even the vilest person who repents.

This is much more difficult for us to accept if we hold Creation in low regard. It is impossible if we consider that we simply evolved over millions or billions of years. In fact, this extreme length of time, in itself, tends to discourage people from having any hope in the possibility of being rescued from our sinful condition. This is one reason some people in some fields of "science" have been led by the evil one to promote the long life of the earth.

There will be a time when God will declare, "Let the one who is doing harm continue to do harm; let the one who is vile continue to be vile; let the one who is righteous continue to live righteously; let the one who is holy continue to be holy."[16] When we have been made holy, nothing but our own choice will change us.

4. God Made the Seventh Day a Memorial.

Only a few blocks from our school in Phnom Penh was another school that had been turned into a memorial. It was the first public display of battered and bullet-riddled skeletons of the killing fields of Pol Pot. Even though the Khmer Rouge were still very active around the country, this reminder to the community was established so the local people would be incited to fight on. The intent was that their fate would not be the same as the bones in the shelves.

What is a memorial, and what does it do? It is usually a place or an activity that serves to remind someone of an important historical event. Most of us will have a memorial stone erected by our family members after our death. We like to remember people we love. A memorial may also be used to focus on a promised future event. There are many memorials found in the Bible. They take different forms. They are not always places.

The Passover was a service God gave Israel. He told them exactly what to do and the reasons why they were to repeat it annually. It consisted of a special meal and storytelling to remind later generations how God had protected them. For the listener, it built faith in God for future protection He would give.[17]

Memorials remind us of things we need to remember. They are sacred to us. What does the memorial of the Sabbath do as a reminder for us? "Then God blessed the seventh day and sanctified it, because in it He rested from all His work which God had created and made."[18] God made the Sabbath a memorial of His work of creating us. He made it a memorial, so we would remember to worship Him as Creator.

THE SABBATH WAS MADE FOR MAN

To whom did God give the Sabbath to? He gave the Sabbath to the same people He gave the earth to. God created the Sabbath for man. God was talking to everyone when He said, "fill the earth and subdue it. Rule over the fish in the sea and the birds in the sky and over every living creature that moves on the ground."[19] He was talking to everyone when He gave the Sabbath to mankind.

Nationalities or religious groups did not exist. No one was separated from anyone else. Adam and Eve represent the entire human race. God gave the Sabbath to them. The earth and its dominion were given to man and all his descendants and so was the Sabbath. The Sabbath was blessed and sanctified for all mankind.

When Moses led the Children of Israel out Egypt and out of slavery, God needed to reacquaint them with the Sabbath. He showed that the Sabbath is a specific day. It is not just any one out of seven. It took this multitude 40 years to make the journey from Egypt to Canaan. The physical trip didn't have to take that long. But, the people needed 40 years to prepare their lives for the Promised Land.

MANNA FACTORY CLOSED ON THE SEVENTH DAY

Early in their journey, the people complained of hunger. They hadn't packed a big enough lunch before they left Egypt! They got hungry and complained to Moses, and he prayed.

God sent manna. The word "manna" means, "What is it?" That is exactly what these desert wanderers said as they watched in disbelief as manna fell on the ground in the morning. "Wow. What is that?"[20] God told the people to collect this unknown substance every morning and eat it the same day. Every day family members gathered enough to feed hungry mouths at home.

Imagine the questions people asked when God told them to only collect enough for the day. Some might wonder how they would know if they gathered enough. Others might try to stow away an extra portion when they thought no own was looking. If a person collected more than enough, their actions showed they did not depend on God for what He had promised to provide.

The women prepared the manna in a variety of ways. It is likely recipes were shared from one tent to another as the women thought of new ways of preparing manna. Any that was not eaten during the day would spoil that night. There were likely feasts taking place right before sundown.

But on Sabbath, God would not send manna. Instead, He followed His own instruction of preparing for the Sabbath ahead of time. The first Friday manna delivery came with a special announcement. The Children of Israel learned that they were to collect a double portion or twice as much as on other days. The Sabbath manna would not spoil during the night. God provided both food and rest for everyone. If they forgot, God didn't make them go hungry on that day. "Those who gathered a lot had nothing left over, and those who gathered only a little had enough."[21]

God didn't randomly choose one day out of seven for us to worship Him. He didn't say, "Pick any day that you want to worship Me. Then on that day you make sure you collect extra on the day before." No, He made a specific day. A specific one of the seven days was singled out. God said, "This is the day on which you are to worship Me. None of the other days will work." After their four hundred years as slaves, He was teaching them what He wanted from them.

But if the Children of Israel went out to gather the manna on the Sabbath, they found the ground was bare. The ones who went out on the seventh day were firmly rebuked.[22] Every Sabbath for 40 years God repeated the lesson. Six days the manna fell. The seventh day there was no manna. The multitude learned to remember the Sabbath. Even today, the Jewish people remember this lesson. They have never missed a single Sabbath since their ancestors trekked through the desert.

REMEMBER! DON'T FORGET!

None of this means we can't or shouldn't worship God on any and every day. It does mean that there is a particular, special day when God asks us to lay aside our other mundane tasks and responsibilities and focus the entire twenty-four hours on Him and our relationship with Him.

Israel learned that God is exacting. He first reminded them when to worship Him through the food they ate. Then, we find it in the Ten Commandments. He wanted to protect mankind from hurting themselves by going against what He had made holy. He said, "Remember the Sabbath day, to keep it holy."[23]

They were already remembering it before this commandment was given. They knew exactly when the manna would fall and when it wouldn't. They already knew which day the Sabbath-day was. God didn't establish it when He gave the Ten Commandments. It was established and made holy at Creation.

There were no Jewish people on Day 7. There were only people who are the first parents of all people on Earth. The Sabbath is therefore given to all people. I was interested to speak with some Jewish believers who had a booth at the New York State Fair. I'm not really sure what they were promoting. I began to speak with one of them about the Sabbath. He seemed upset that a non-Jew would be interested in the Sabbath. He told me it is only for them. I have a great appreciation for people of the Jewish religion. Still, no one has more exclusive claim to the Sabbath than they do to grass. "The Sabbath was made for man."

SIX DAY WORK WEEK APPROVED BY GOD

"Six days you shall labor and do all your work, but the seventh day is the Sabbath of the Lord your God. In it you shall do no work."[24] Notice that it speaks of six days of labor. God intended man to work six days out of the week. After six days of hard work, you can be sure that by the time the seventh day comes around, man is ready to rest. The six days of work were also made for man. As a result, the seventh day is even more of a blessing because of the work of the previous six days.

We have looked at many concepts God originated during Creation week that have been changed by man. Perhaps it has been to downplay the likelihood of God's existence. When we reduce the number of days in the work week from six days to five (and in some places four days), do we diminish our need for rest on the seventh day? When people are confused as to when the seventh day is, how much of the reason is to take God out of the picture?

But the commandment continues and tells us who should rest on the Sabbath day and who should not work. God included everybody. No one is to work on your property. "On that day no one in your household may do any work. This includes you, your sons and daughters, your male and female servants, your livestock, and any foreigners living among you. For in six days the Lord made the heavens, the earth, the sea, and everything in them; but on the seventh day he rested. That is why the Lord blessed the Sabbath day and set it apart as holy."[25]

This commandment even tells us that it is given so the world will remember that God is Creator! The Sabbath is a memorial or reminder of Creation. (That is what "remember" means.)

As such, one purpose the Sabbath has is to remind man that God is Creator, and the world was made by Him. Imagine the implications if man never forgot the Sabbath or turned away from it. God would never have been forgotten as Creator. How different the world would be today if man had always remembered that God is Creator and worshipped Him.

We have been warned that before Jesus returns, man will fail to worship God as Creator, with words that come from the fourth commandment—pointing us back to the Sabbath. "Worship Him who made heaven and earth, the sea and the springs of waters."[26]

THE SABBATH REST REMAINS

Does the Sabbath rest remain important in the New Testament? The writer of the letter to the Hebrews believed it does. "For somewhere he has spoken about the seventh day in these words: 'On the seventh day God rested from all his works'…Therefore since it still remains for some to enter that rest, and since those who formerly had the good news proclaimed unto them did not go in because of their disobedience, there remains, then, a Sabbath-rest for the people of God; for anyone who enters God's rest also rests from their works, just as God did from His. Let us, therefore, make every effort to enter that rest, so that no one will perish by following their example of disobedience."[27]

The Sabbath is our portal to spending an eternity with our Creator. Keeping it does not save us, but through the Sabbath we come closer to God and are drawn to surrender to Him. God longs for us to understand why He created the Sabbath. This day above any other will prepare us to avoid the beast who seeks as a roaring lion to destroy us.[28] God's greatest desire is that we learn to find rest in Him in this special portal of time. How will you spend this day?

Chapter 12

CREATION: AFTER THE FALL

I had heard of this room. It was one of those places you would love to spend time in if you are interested in the behind-the-scenes activities of national and global crises. The Situation Room of the Department of State in Washington D.C., had probably dealt with situations I lived through only a few years previously in Cambodia. Now, our world Director[1] had arranged a private tour of the site. It was a chance for me to receive some insights into the conditions I would be moving into, and to ask questions. I was traveling to Thailand to work with refugees arriving from Vietnam, Cambodia, Laos, and Burma. It was clear to see many people were interested in what was going on at the Thai border. Our work would directly affect the lives of tens of thousands of people escaping from the beast and his minions.

HEAVEN'S SITUATION ROOM

So, pull back the curtains—not of time, but the curtains of a different dimension. Look into the very Situation Room of Heaven to see the rescue mission God established for you. It displays His great love for each one of us.

When God decided to make Earth and Man as the crowning act of universal Creation, He didn't make us without thinking and planning. God didn't just arrive at the day He began creating and say, "I think I will make an earth! Let Me just throw some stuff here and see what comes up." No, the creation of earth and man was a part of God's intention since He began creating the universe, which had been happening for eternity.

At some point before the beginning of earth's creation, God fashioned a mighty being who He named Lucifer. The Bible describes Lucifer as a powerful angel. "How you are fallen from heaven, O shining star, son of the morning. You have been thrown down to the earth, you who destroyed the nations of the world."[2]

"For you said to yourself, 'I will ascend to heaven and set my throne above God's stars. I will preside on the mountain of the gods...I will climb to the highest heavens and be like the Most High.' Instead, you will be brought down to the place of the dead, down to its lowest depths."[3]

Lucifer believed he should sit on the throne of God. He thought a lot of himself. He had reason to, except there was nothing he could do that God had not made possible. It is the same as with any of us.

The Bible tells us more about this beautiful, mighty angel. "You were the model of perfection, full of wisdom and exquisite in beauty. You were in Eden, the garden of God. Your clothing was adorned with every precious stone...all beautifully crafted for you and set in the finest gold. They were given to you on the day you were created. I ordained and anointed you as the mighty angelic guardian. You had access to the holy mountain of God...You were blameless in all you did from the day you were created until the day evil was found in you."[4]

What does it mean to be the angelic guardian? Lucifer's position in heaven's hierarchy placed him right next to the throne of God. After God Himself, the Father, Son, and Holy Spirit, Lucifer was the most powerful being, the most powerful angel in the universe. Yet he began to look at God's throne and think, "I could do that. I could take the place of God." That is where he began to stumble.

Lucifer circulated throughout Heaven conspiring with other angels. Although he did this discretely, God always knew what was happening. Lucifer whispered to the other angels, "You know, if I were on that throne, I would do wonderful things for you." He began to plant seeds of doubt about God in the angels' minds. "God is just not fair! He says He is a loving God. But if He were loving, would He do this? Would He do that?"

IF GOD IS A GOD OF LOVE, WHY DOES HE DO THAT?

Today many people say the same thing. They question God's love and fairness. Lucifer's influence began to foment dissatisfaction. Many of the angels became uncomfortable with God's rule and dominion. Lucifer's skillful tactics led angels to rebel against God.

If you were God, what would you do? You are in charge of the entire universe. You possess the power to do anything You choose. You can make anything or destroy anything by merely speaking the words. In fact, all You must do is think the action, and it will occur.

Rather than simply destroy the errant one and remove the memory from every mind, God allowed the unthinkable. There was war in Heaven. What a somber thought. Those two words don't seem to go together—war and Heaven. But that is what the Bible tells us. "Then war broke out in Heaven. Michael and his angels fought against the dragon, and the dragon and his angels fought back. But he was not strong enough, and they lost their place in Heaven."[5]

The devil was just not strong enough. Didn't he have enough angels on his side? No. But no matter how many angels Lucifer rallied to his cause, the numbers would never be adequate. God is stronger than any number of angels. Lucifer and his angels lost their place in Heaven because of the choice they made to separate themselves from the God who formed them.

Thus, the beast entered the Earth stage within the confinement of the Time Box. "The great dragon was hurled down—that ancient serpent called the devil, or Satan, who leads the whole world astray. He was hurled to the earth and his angels with him."[6] He had many names. The dragon, the devil, the serpent, Lucifer, Satan, shining star, son of the morning—the beast.

How many of the angels were cast to the earth? "Its tail swept a third of the stars out of the sky and flung them to the earth."[7] Lucifer took one-third of the stars or one-third of the angels, and they were cast here to the earth with him.

This created a huge issue for God, for all the earth, for you, and for me. It created a predicament that man cannot solve. With His ability to look forward, God had known Satan and the angels would rebel. There is no such thing as time for God. Time does not constrain Him. God can travel throughout the time continuum in the same way you and I can get into a Volkswagen and drive from New York to Tennessee. God can get there much faster than the Volkswagen. Time does not hold God captive.

God looked forward and envisioned the extraordinary beings that He wanted to make. He saw a world full of people who loved Him. God longed to shower His love upon these cherished people. With Satan's actions, God's heart ached. He knew the result of the heavenly rebellion would cause humans to sin.

GOD'S HEART ACHED

God knew Lucifer would be responsible for causing the eternal death of many. In fact, Satan would claim all humans if God did not have a solution to the sin problem. For God there was only one way to solve the problem of sin. He saw it would not be to simply destroy everybody who has gone against Him. There would be no one left on Earth.

God's plan was personal. It was expensive. It was painful. He would allow His own Son, Jesus, to create the new world. Then, after mankind sinned, Jesus would be born and live among the very people He created and loved to provide a way of salvation from eternal death and give evidence of just how much He loves us.

The evidence of His foreknowledge and planning is in the water and land and light of Day 1 and the atmosphere of Day 2. It can be seen in the seeds of all the plants of Day 3 and the sun and moon of Day 4. It becomes obvious as we look back at life and birth on Day 5 and man himself on Day 6. But God didn't stop there.

He knew in advance who would accept Jesus and the salvation offered. He also knew who would follow the lead of Lucifer and rebel against Him. "They are the ones whose names were not written in the Book of Life that belongs to the Lamb who was slaughtered before the world was made."[8]

Everyone who chooses not to follow Jesus—the Lamb of God, will follow Satan. The scenario of choice, which occurred in heaven with the angels, played itself out once again on Planet Earth. Revelation 13:8 tells us a crucial fact. If your name is written in the Lamb's Book of Life, it is because you have chosen to serve Jesus and not to follow the evil one. There must be a conscious choice to follow God.

This is God's plan. And it would all be wrapped in the box of time. Beyond the bounds of the box of time, sin could not travel.

Before sin and the rebellion in heaven, God already knew if He created this earth, He would send His Son to die for creatures He would make—for us. God the Father and God the Son decided to go ahead. Then they did something no one would have expected.

GOD ESTABLISHED A WAY OF REBELLION

What a thing for God to do! God wanted us to love Him. This love needed to be real and freely given. We could not be forced to love Him because we have no options. He wants us to enjoy being with Him. If you are forced against your will to be with someone all through your life, the relationship is repugnant. If God expected to us to be with Him for eternity out of fear or obligation, eternity would be hell for us rather than heaven.

God planted a tree in the Garden of Eden and named it the Tree of the Knowledge of Good and Evil. It may seem strange that the name included the word or idea of "Good." After all, for those who know about that tree, what we know certainly isn't good. It is easy to see that "Evil" is in the name since it was used by Satan to bring sin to the earth and all its inhabitants. That is about as far from "good" as we can imagine.

Adam and Eve knew nothing about evil. They had only one condition to live in perpetually, until sin. That condition was "good." God presents the concept that there is something else other than what He had preferred to have made. Had He hidden the tree, Lucifer's accusations would have included that as evidence of unfairness.

The term "good" covered everything God created. At the end of each day, He said it was all good. This term wasn't a name, but a description. Everything was good. God had already experienced good and evil. He knew of it before it developed in the heart of Lucifer. Then God had fought the battle with evil in the very place that no one would ever consider evil might be found—in heaven.

In Genesis 3, we learn that sin came into the world. We have no idea how long Adam and Eve were in the Garden before sin began. I like to think they were able to enjoy the perfection of God's creation for a long time before they finally fell to the temptation to sin. But the Bible doesn't tell us.

ONLY TWO CHOICES

We do know how it happened. Genesis 3 says the devil took the form of, or simply spoke through a serpent. The evil one waited at the Tree of the Knowledge of Good and Evil. It is the only place God allowed Satan to be on earth. Why? There had to be opportunity for mankind to turn from God if we choose to do so, but He wanted us to live free from sin. In our own lives, sin and evil surround us, but through our choices, we can avoid many places and situations where temptation is greater. God didn't want humans to be challenged with sin all the time. So often we make conscious decisions to place ourselves on the devil's ground. God told Adam and Eve they were to stay away from that one tree.

"Then the Lord God took the man and put him in the garden of Eden to tend and keep it. And the Lord God commanded the man, saying, 'Of every tree of the garden you may freely eat; but of the tree of the knowledge of good and evil you shall not eat, for in the day that you eat of it you shall surely die.'"[9]

One restriction seems like an easy enough test. Throughout the garden, other trees hung full of beautiful, luscious fruit. Adam and Eve didn't need to go to that particular tree. Many have asked, "Why

176 • GARY WAGNER

did God make such an arbitrary, unnecessary rule?" It is because there had to be freedom to choose. For God not to have given opportunity to choose would have made us slaves. We wouldn't even have known we were slaves.

A PRISON IN PARADISE

For several years, my family and I lived in idyllic Pacific Island settings. They are the picturesque locations you see in travel magazines. They beckon you to sit on the beach forever or snorkel among the coral and schools of tropical fish. People spend a lot of money for a two-week vacation there.

If you live on the island, it is another story. Once you step off the plane and settle into life's routine, a curious phenomenon begins to form within you. The sandy beaches, the swaying palm trees, the lapping of the lagoon water take a toll. Island fever starts to set in. You just want to get off the island. You are tired of seeing water everywhere you look. You begin to hate the beautiful place. If you can't get off for a short break, you feel trapped, like a prisoner.

Adam and Eve needed to know they had options. God desired for them to forever see the Garden as beautiful, wonderful. After all, God was there. But, God had to offer a choice. Yes, the options may seem harsh. "I want you to live with Me. I want you to be with Me. I want to enjoy your company. I want you to enjoy My company. I am the One who gives you life. My presence here with you makes it possible for you to live. I have given you this gift of life. But I have also given you the power of choice.

"If you choose, you can turn away from Me. You are free to make your own choice. You can decide whatever you want. If you don't live according to the laws that give life—you intentionally move away from the life that is given." This choice is an issue of eternal life or eternal death.

The reality is, there are only two choices in life. We are talking basic life. Not the supermarket-shelf choices. Each time I return from living in a developing country for an extended period, my first trip to the supermarket brings the same reaction. I walk through the doors and go to the first aisle. My mouth drops open as I stare at all the options, the incredible abundance. That is not the kind of choices we are talking about here.

DID PEOPLE LONG AGO HAVE IT BETTER?

In Noah's day, people either went into the ark or stayed outside.[10] There were only two choices. In Moses' day, people either put the blood on their doorpost, or they did not.[11] There were only two choices.

Jesus told parables about the wheat and the tares,[12] the sheep and the goats.[13] Each time there were only two choices. God asks us to choose life. The wonder is, the promise is, if we choose life, we will have it!

Each of us is making this choice with every decision we make. We may or may not recognize the eternal consequence of every decision, but it exists. We need to become aware and consciously make our choice.

In the end, our choices will be exhibited by one or the other of two options for our eternity. People receive either the Seal of God[14] or the mark of the beast.[15] Just two choices. In the final judgment, there are just two classes...the just or the unjust. And when the New Jerusalem comes down from God out of Heaven, and everyone meets for the first and last time, there are just two classes. Those who are on the inside and those who are on the outside.

178 • GARY WAGNER

God wanted Adam and Eve to understand completely the cost of separating themselves from the Life-Giver. The devil was waiting at the tree and spoke to Eve through a snake. It was the first time humanity was confronted with the choice.

"Then the serpent said to the woman, 'You will not surely die. For God knows that in the day you eat of it your eyes will be opened, and you will be like God, knowing good and evil.'"[16]

That sounds pretty good, doesn't it? Eve surely loved her Creator. The possibility to be like Him was certainly an enticement. Some of what Satan said was true. But God had clearly said not to eat of that tree.[17] Eve decided to trust the words of the serpent more than those of God.

"The woman was convinced. She saw that the tree was beautiful, and its fruit looked delicious, and she wanted the wisdom it would give her. So she took some of the fruit and ate it. Then she gave some to her husband, who was with her, and he ate it, too. At that moment their eyes were opened, and they suddenly felt shame at their nakedness. So they sewed fig leaves together to cover themselves."[18]

A CURSE, A CONSEQUENCE, A PROMISE

Adam and Eve didn't die the day they ate the fruit. Why? Was God's pronouncement a lie? No. The curse was placed, but not on man. After Adam and Eve sinned, God came looking for them. I believe it was Jesus. Three sets of consequences were administered.

"Then the Lord God said to the serpent: 'Because you have done this, you are cursed more than all the animals, domestic and wild. You will crawl on your belly, groveling in the dust as long as you live.'"[19] The serpent was cursed. But God didn't stop with the physical curse upon the serpent. There is more to the curse.

"And I will cause hostility between you and the woman, and between your offspring and her offspring. He will strike your head, and you will strike his heel."[20]

In this part of the curse, God was addressing the devil and not the serpent. The beast had attacked man directly. It proves his hatred for us. The main part of the curse had nothing to do with nature, but with the one who had usurped a part of nature to do his own bidding.

The fact that this part of the curse was directed at Satan personally was more than just to receive deserved eradication. It was also to humiliate him. The evil one touched the two most beloved creatures of God, and with them, everyone who would be born as long as time would last. To crush the head of the serpent would be to destroy him. It would not be God who would execute the annihilation of Satan, but a Man. It would be the One who would come and be born God with us.[21]

In these words, God spoke the promise of the Savior.

God knew that Adam and Eve ate from the forbidden tree, even before they came out of their hiding place. Their sin caused them to be ashamed of their nakedness, so they made coverings for themselves. Sin invaded their lives, and they hid from God in fear.[22]

We have all had that sinking feeling of knowing we have done something that would change our lives for the worse. But none of us has ever had this deep of a sense of foreboding. To go in an instant from perfect innocence and oblivion of evil to the recognition of one's own absolute guilt and fear of deserved retribution had to be utterly terrifying.

But in His great love and compassion, God came tenderly to their side. He could not simply wipe away the deed or its penalty, but He could grieve with them for what it would cost the world and Him.

He could say to them, "A Rescuer, a Savior will come for you."

Chapter 13

WHAT'S A SAVIOR TO DO?

Jesus spoke to Adam and Eve about the natural consequences of their decisions to turn against Him. "To the woman He said, 'I will make your pains in childbearing very severe; with painful labor you will give birth to children. Your desire will be for your husband, and he will rule over you.'"[1]

God's words to Eve were not a curse. When God places a curse, He says it is a curse. Look at the words God spoke to the serpent. He stated clearly that the serpent was cursed. When God spoke to Eve, He said two things would happen. These things were for the sake of the woman and not as a curse upon her. One is pain in childbirth. Struggles and pain are allowed to continue to direct our eyes to God—the Healer and the Savior.

Think of the time of any great national catastrophe. Whenever we hurt in masses, there is an overwhelming tendency for many to say, "Oh God. Please come and help us. Please bring us through this terrible time." I watched it happen in Cambodia during the time of Pol Pot, and in the Republic of Congo in 1993 during the beginning of a civil war. Pain, misery, and difficulty cause us to turn our eyes back to God. We see it every time there is an attack in our schools and other public places by terrorism and derangement.

In childbirth God wants a woman to focus on Him. He wants her to remember that this gift of birth comes from Him. The pain is to point her back to the fact that the Creator gives life. It also serves to make us yearn for the promised end to suffering when the Savior will come.

Eve strayed from Adam and was beguiled to sin. Her natural interest in seeing the wonders that God had made led her to exhibit greater independence and move into areas of life she was unprepared to face on her own. The consequence for her was that she would have a desire to be with her husband to protect her from these perils. In many ways, this is a very positive result. It is not that women cannot be independent, but that they also have the provision for protection.

A husband must be chosen who will be this protector. It is important we recognize the situation of Eve. A woman anticipating her husband-to-be must be sure he is of high moral character to be trusted with her life. This is not to give husbands permission to hold power over their wives, but to give them physical, spiritual, emotional, and social cover when they need it.

How often I have heard, "But there just aren't that many good men like that anymore." This is what Michael Hyatt calls a limiting belief. We see a real or imagined obstacle in our path and decide we can't get through it. That belief will prevent us from overcoming the obstacle.[2] We become impatient and settle for something less than the optimum. I have seen that if good women insisted on finding good men, and refuse to accept a lesser character, there would be more good men. This is not to blame women for the lack of good men. It merely shows the power of women to affect the direction of society.

There is also a message for the men. "Then to Adam He said, 'Because you have heeded the voice of your wife and have eaten from the tree of which I commanded you, saying, 'You shall not eat of it':

Cursed is the ground for your sake; In toil you shall eat of it all the days of your life.'" [3]

Adam knew that what he was doing was very wrong. His reasons do not matter for this work. Still, the curse was not on him.

Both the man and the woman would endure hard labor. For both, the purpose was to turn their hearts to God for relief. Later, Jesus would make the reason clear. "Come to Me, all you who labor and are heavy laden, and I will give you rest. Take My yoke upon you and learn from Me, for I am gentle and lowly in heart, and you will find rest for your souls." [4] This tender sentiment was the same with which He approached the frightened Adam and Eve in the Garden. It is the way He approaches us to invite us to give our burdens to Him.

THE CURSE WAS ON THE GROUND

Adam was not cursed. Humankind was not cursed. The ground was cursed. It says, "Cursed is the ground for *your sake*." (emphasis supplied) The soil became different. "It will grow thorns and thistles for you, though you will eat of its grains. By the sweat of your brow will you have food to eat until you return to the ground" [5] We will see later that this is a change at the genetic level.

Man was to toil. He was going to work hard. Taking care of the garden would no longer be just a pleasant stroll through it. It was to be hard work, "the sweat of your brow" and "thorns and thistles." This struggle is on a daily basis for those six days every week when we are to work. The purpose is to point man back to God.

It was also to remind us to depend on Him for rest from the labor. It is the reason God instituted the Sabbath on Day 7. He described it as a day He rested from His work. [6] Later, when He was rebuilding His chosen people, He told them, "Remember the Sabbath day by keeping it holy. Six days you shall labor and do all your work, but the seventh day is a Sabbath to the Lord your God. On it you shall not do any work."[7]

This is another example of how God provided something during Creation before it was seen to be needed. Knowing man would sin and that a consequence would be labor, He put the Sabbath in place to provide needed rejuvenation to accomplish our work.

As a natural result of the curse on the ground, it was to be a hardship to make a living and to grow the crops needed for life. This pain, too, was to draw man to God. The person who recognizes the cause of our hardships and the purpose for them will hopefully appreciate that they are bringing us to God. As we surrender to Him, He can prepare us for eternity.

The verse also reminds us that the looming consequence of sin is death. It was what God had foretold, and what He had sought to prevent by His warning.

Many people look at death as a curse. It is not. We will see a little later that death is a natural consequence and may actually be a blessing in this life. This discussion brings us to an extremely crucial question. Why did Adam and Eve not die that day?

JESUS TOOK THE CURSE FROM THE GROUND

Sinless Jesus became sin for us when He hung on the tree. Remember when we studied Day 3 we learned how Jesus made the beautiful tree that His creatures would use to kill Him. "Christ has rescued us from the curse pronounced by the law. Crucified on the cross, Jesus took upon Himself the curse for our wrongdoing. For it is written in the Scripture, 'Cursed is everyone who is hung on a tree.'"[8] Jesus took the curse for us, but He did even more than that.

Throughout the Bible, we find evidence of the natural consequences of sin. Sometimes, the consequence is immediate death. In the long run, it always is the final result.

The Bible gives examples of valid reasons for a death penalty. Even when there was such a penalty imposed, there were rules about the process. One such law speaks of death by hanging on a tree. "If someone has committed a crime worthy of death and is executed and hung on a tree, the body must not remain hanging from the tree overnight. You must bury the body that same day, for anyone who is hung is cursed in the sight of God. In this way, you will prevent the defilement of the land the Lord your God is giving you as your special possession."[9]

Sin and righteousness are not something you do, but something you are. God is righteous. He doesn't do righteous. To forgive man's sin, Jesus became unrighteousness. He became sin.

You cannot be two opposing things at once. You cannot be righteous and a sinner. Sin cannot exist in the same being as righteousness. On the cross, God "made him who had no sin to be sin for us, that in Him we might become the righteousness of God."[10]

Remember that the ground was cursed for our sake. Now Jesus became sin for us. He didn't do sin. He didn't commit sin. He became sin. He became cursed when He was hung on the cross. "Cursed is everyone who hangs on a tree."[11] This intricate plan was orchestrated to cover 4,000 to 6,000 years and involve each person who lived. Everyone has been touched. No one is excluded by God from the right to make our own choice.

"When Adam sinned, sin entered the world. Adam's sin brought death, so death spread to everyone, for everyone sinned."[12] Sin became a part of our very nature. When we are honest with ourselves, we know we also commit sin.

Sin is a condition of being separated from God by some barrier. The barrier is death brought on by sin. Jesus did not have the benefit of being able to be sin, and at the same time, be declared righteous. The sin that He became killed Him. He allowed Himself to be made sin for us. "Sin is the sting that results in death, and the law gives sin its power. But thank God! He gives us victory over sin and death through our Lord Jesus Christ."[13]

IN THAT VERY DAY

The time box fulfilled a part of its purpose here. As has been mentioned, God is not limited by time. It is a dimension He can pass through at will. Jesus "was slain from the creation of the world,"[14] to show the universe that He knew what would happen. His decision to rescue mankind was made before there was a need. It was not an afterthought. Jesus moved forward in time before the fruit was taken from the tree.

God put the curse on the ground in Genesis 3 for us. Jesus was grafted onto a tree with spikes and became a branch. The tree was planted firmly in the ground. The curse of sin that was in the ground came through the tree and into the branch which Jesus was, and He became sin for us. It was not because He did anything that was wrong, but because He accepted that only His sacrificial death could rescue mankind. He agreed to be grafted into the tree. Jesus died that very day of His crucifixion in fulfillment of the admonition, "…in the day that you eat of [the tree of the knowledge of good and evil] you shall surely die."[15]

Who was cursed by the sin of Adam and Eve? Was it the woman? No. Was it the man? No. The ground? Yes. The serpent? Yes. And Jesus was cursed. Jesus was cursed because of the sin of the man and the woman and His willingness to become the curse. What does that mean for our sin? When we sin what is the effect on Jesus? He is cursed because of our sin. We say we don't take His name in vain, but when we sin we are cursing Jesus.

If God had not previously taken the curse upon Himself, Adam and Eve would have died the day they ate the fruit. The fact that they did not die that day shows us that in light of eternity Jesus had already been slain for their sin.

After Jesus died, the ground claimed its prize of the One who was sin. He was buried. But He overcame death and burst from the tomb. When His followers went to the grave on the third day, they were met by an angel who asked them, "Why do you look for the living among the dead? He is not here; he has risen!"[16] The curse on the ground had no more power over Him. Even though He paid the price for sin and had been returned to the ground, He had victory over death when He arose from the tomb. The curse was defeated for Jesus, and for every person who accepts Him by faith.

Jesus could do this because He was the Creator. He was the Law-Giver. Only He could pay the price for the law which He had given. By doing so, He fulfilled the Old Testament prophecy of Hosea. "I will ransom them from the power of the grave; I will redeem them from death. O Death, I will be your plagues! O Grave, I will be your destruction!"[17]

The curse has been defeated on our behalf. It is clear that Jesus even died for those who do not accept Him. Had He not died and risen, Adam and Eve would have died when they ate the fruit. It would have been the end of Creation. But He did rise. This gives the unrepentant soul time to consider God. They, too owe their lives to Jesus.

After the resurrection of Jesus, He returned to the time of man's sin. I believe that when Jesus came to see Adam and Eve in the garden asking, "What have you done?" He already bore the scars in His hands, side, and feet. They wouldn't have recognized that. They were consumed by their guilt and fear. They wouldn't have understood what the scars were.

When He spoke to the serpent saying, "there will be one who will come who will crush your head and you will crush his heel,"[18] it was because He had already been there. He knew it would happen because He had already done it in the future.

So, because Jesus had already paid the price of their sin on the day that He became sin, Adam and Eve didn't die that day.

THE EFFECT OF THE CURSE ON THE GROUND

"Against its will, all creation was subjected to God's curse. But with eager hope, the creation looks forward to the day when it will join God's children in glorious freedom from death and decay. For we know that all creation has been groaning as in the pains of childbirth right up to the present time."[19]

Every day of life is a day of grace. By rights, we should die for every sin. The multitude of people on the earth is evidence of the magnified grace of God. Even with that grace, sin takes its toll on our physical bodies. We no longer have access to the Tree of Life. We are living a slow death. That can be ameliorated by maintaining the rules established in the beginning. Health comes from the diet God made and mentioned on Day 6.[20] Strength comes from the prescription of work by the sweat of your brow.[21]

There is still another result of the curse on the ground. It is the deterioration of all life. We are thankful for scientific advances to provide evidence of this biblical fulfillment.

ANOTHER RESULT OF THE CURSE ON THE GROUND

It is crucial that we remember that Adam was made from the dust of the ground.[22] When we die, we will return to dust. We will always have a powerful connection to the elements that comprise every part of our bodies. All life has this same unbreakable link. What happens to the earth happens to us. This action also refers to the reality of the breakdown of our tiniest building blocks.

Professor John Sanford of Cornell University tells us that the human race is degenerating. According to his research on genetic entropy, all of creation is slowly deteriorating. "[Life] is not an upward spiral of evolution, where things keep getting better and better. It is a downward spiral exactly as described in Scripture."[23]

Dr. Sanford says, since the beginning (I would suggest since the beginning of sin), "We have about three mutations [for] every cell division in our body...We are adding about three new mutations per cell per day."[24]

On another level, "I take the genes, all the mutations I inherited from my ancestors, which is tens of thousands of deleterious mutations in my body, and then I add my own contribution to that by a hundred new mutations at least, and pass it on to the next generation."[25]

Sanford also says, "Personal genetic entropy is an uncontestable fact that no scientist on the planet can deny because it is why we die."[26] His conclusion is, "We are a perishing people living in a dying world."[27]

Michael Lynch is among the world's top population geneticists. He sees the same mutation process happening to all life. His conclusion, however, does not support a Creation mindset. Even though he finds his explanations in evolution, the conclusions seem to be basically the same. According to Lynch, "…it remains difficult to escape the conclusion that numerous physical and psychological attributes are likely to slowly deteriorate in technologically advanced societies…on a timescale of a few generations, i.e., 100 years."[28] Lynch later includes intellectual and behavioral attributes. This means that within the next three to four generations, possibly one hundred years or less, mutations will cause a noticeable decline in our functional physical and mental attributes.[29]

This decline is happening everywhere, but especially in places like the United States and the United Kingdom where there has been a slow decline in intelligence over the past century.[30] How can these things be in countries that have possibly the best medical care in the world?

It may be partly because of this advanced medical care. "In the United States, the incidences of a variety of afflictions including autism, male infertility, asthma, immune-system disorders, diabetes, etc., already exhibit increases."[31] Some heart disease, metabolic diseases, psychiatric disorders, and cancer are other additions to this list. At least a part of this can be credited to medical treatment giving extended life to people carrying mutated genes. This can allow the malady and an increased number of mutations to be passed on to others.[32]

HUMAN EXTINCTION

Evolutionary biologist Brian Charlesworth[33] considered the number and rate of mutations. He seems to have difficulty fitting the evidence into his allegiance to evolution. He titled an often-quoted paper "Why We Are Not Dead One Hundred Times Over."[34] Sanford reinforces this. "In deep geological time we should have been extinct a long time ago."[35] In other words, if life had been around for millions of years as some evolutionists say, then with the mutation rate we experience, the human race would have already wholly died out a hundred times over by now! Genetic entropy disproves evolution. It shows evolution cannot possibly explain the development or the continuity of life.

All of this refers only to the mutations that occur in the DNA of every person. On another level, all our cells are deteriorating. This is what causes aging. "By the time we are 60, we have tens of thousands of new mutations per cell. This is what limits our life expectancy, and no medical breakthrough can be expected to halt continuous mutation in virtually every cell of our body."[36]

This news brings anxiety to the hearts of evolutionists who believed that natural selection is making everything better. There is no need to fear. Through Jesus, we are freed from sin and its curse. "For when we died with Christ we were set free from the power of sin."[37]

We are brought full-circle. We were created by God in His image, after His likeness.[38] We fell from that lofty position. But God will return us to be in His image. "If anyone is in Christ he is a new creation."[39] Now, by surrendering to Him, we may choose to allow Him to re-create us in His image. He will change everything we are willing to give up to Him. For most of us, this will happen by degrees.

He changes one thing, we see the blessing of what He has done. We trust Him with continually more of our lives. As we allow Him to work, He will completely renew His image in us. Here is the secret code that gives us entry to the rescue vessel to eternity. His image in us is everything. This same process will prepare us to receive new bodies. "For our dying bodies must be transformed into bodies that will never die; our mortal bodies must be transformed into immortal bodies."[40]

It is our key to entry into eternity, and it is all by His grace.

None of this change is accomplished by our works or good deeds. Only God can remake us in His image. Only He is good. Only He has the power and the intellect. Only He has the authority, because only He was perfect and paid the penalty for our sins by His own death. And He is waiting for you to give Him the invitation to begin this work in you. Why not now?

For now, evil continues to stalk us. The ground still bears the curse of sin. Entropy marches on and engulfs each succeeding generation. When the body contracts the ailments of the ground, cursed because of our sin, we eventually succumb to the need of rest.

INSTRUMENT OF GRACE

Every day of life is a day of grace. Death comes to us as an instrument of the grace of God. How cruel it would be, if we were forced to continue to live in misery and pain for millennia, waiting for the promise of a return to glory and righteousness. Rather, we can be laid to rest to wait for the resurrection. That is a gift of grace, an escape from the misery and pain brought on by the sin of the world as well as our own. Death is not a punishment, but rather the natural consequences of being in a world dominated by sin.

Even Adam must have felt this way after watching the decline of humanity for nearly a thousand years. "Though the sentence of death pronounced upon [Adam] by His Maker had at first appeared terrible, yet after beholding for nearly a thousand years the results of sin, he felt that it was merciful in God to bring to an end a life of suffering and sorrow."[41]

DEATH ITSELF WILL BE DESTROYED

The dying world is the natural result of the devil's character and style of leadership. But Jesus through His death and resurrection gained the victory over death. "Then the end will come, when he hands over the kingdom to God the Father after he has destroyed all dominion, authority and power. For he must reign until he has put all his enemies under his feet. The last enemy to be destroyed is death."[42]

The Bible explicitly describes the destruction of death. We don't have to worry about a slow death by ever-increasing mutations. Instead, we look forward to the return of Jesus, and a time of judgment when every soul receives the desire of their hearts. Some want to be with God. They have waited for Him, and He will usher them into the heavenly kingdom. Those who chose not to be part of God's eternal plan will receive their wish. Not to be with God will mean not to be at all. God is everywhere, so they cannot be anywhere. He will not continue to allow rebellion to affect His Creation the way sin has affected Earth and its people. None of us want that kind of existence.

How will those who do not choose God meet their final end? "And fire came down from God out of heaven and devoured [the wicked]. The devil, who deceived them, was cast into the lake of fire and brimstone...Then Death and Hades were cast into the lake of fire. This is the second death."[43, 44] It is the death that puts an end to death. It is the fulfillment of the promise that "Death is swallowed up in victory."[45, 46]

There will be no corner of the universe where unfortunate or malevolent souls languish for eternity in suffering and misery. For there to be such a place would be to contradict the promise of God. He will "wipe every tear from their eyes, and there will be no more death or sorrow or crying or pain. All these things are gone forever."[47] This verse does not talk about anyone continuing to live forever in misery and punishment.[48]

Each of us has the freedom to choose life or death. This is eternal life or eternal death we are talking about. He wants you to be with Him throughout eternity in a condition of perfection, righteousness, joy, and happiness. He wants you to experience life with no pain, no sin, no suffering, no tears. This was His plan when He made us in His image. This is what He calls us to. This is why He sent His Son to rescue us.

Chapter 14

TWO GARDENS AND THREE TREES

God instructed man to "fill the earth and subdue it; have dominion over the fish of the sea, over the birds of the air, and over every living thing that moves on the earth."[1]

Throughout the narrative of Creation, at the end of each day, God reviewed what He had made. Each day His evaluation was, "It is good." Nothing that existed was not good because God made it. He made it wonderfully, and He made it to be perfect.

Life on earth was intended to last forever. Thorns or splinters, illness or weakness, sorrow or sadness were not supposed to exist. Death was not even imagined. Lack of these life-limiting factors is part of the description of paradise.[2] We cannot comprehend the glory and beauty of the earth that God gifted to man. In this perfect setting, man had every physical and spiritual need met. After the creation of Eve, the first couple could not even imagine anything that they needed or would enjoy that they did not already possess. Until the telemarketer called.

The eternal plan for man in paradise included more than the beautiful Garden. God knew the exact decisions the man and woman would make, and He ordained every commitment and provision for eternal happiness despite those wrong decisions. The big picture of eternity contains three phases of which we have been made aware.

The first is Creation itself and placing man in the idyllic Garden of Eden. Everything man needed to exist was there. The beauty of the earth, the necessities and joys of life, and companionship. Love reigned supreme. The metanarrative expands eternally beyond these seven days. It begins here but moves through a period of cataclysm and returns to a second garden.

The second garden already exists in the endlessness of time without time. We will not see it until the drama of all ages is completed here on Earth. Many call this new garden Heaven or Paradise. Under different names, it is the anticipated goal of nearly all world religions. In this place will be the tree found in Eden. It is the Tree of Life. Of the three trees discussed in this chapter, it is the only one that is located in both gardens. It stands before and after sin, as a promise of fulfillment of the abiding love of God and the already present promise of endless joy.

However, two more trees factor into this narrative. Their roles are imperative. They explain how we get from one garden to the other. The second tree is the Tree of the Knowledge of Good and Evil. This tree was the evidence of our freedom, but the symbol of our enslavement to misery and sin. Because sin occurred, we must understand it to reach the second garden. But more than understanding is needed.

The third is the tree upon which God and the Son of God paid the penalty for our sin. It is the Cross of Christ. It is not found in either garden, but the promise of it is discovered in Eden. Its fulfillment will be seen in Heaven. Accepting the sacrifice made on it will take us to the second garden. Ponder the trees.

THE TREE OF LIFE

The Tree of Life is located in two different settings. First seen in the Garden of Eden, it drops out of sight after sin enters the picture. We see it again in Heaven. Each time the function is the same.

God placed the Tree of Life in the middle of the garden.[3] This location demonstrated the importance of the tree, particularly to man. Other plants provided energy and sustenance for everyday life. The fruit of the Tree of Life was unique. If Adam and Eve ate "fruit from the tree of life…they [would] live forever!"[4]

In some ways, we still have access to a tree of life. Proverbs tells us "Wisdom is a tree of life to those who embrace her; happy are those who hold her tightly. By wisdom the Lord founded the earth; by understanding he created the heavens."[5] Other principles described as a tree of life to us include the "seeds of good deeds" (they lead to winning friends),[6] "fulfilled dreams,"[7] and "gentle words."[8]

These gifts are available to us now. "If any of you lacks wisdom, let him ask of God, who gives to all liberally and without reproach, and it will be given to him."[9] He has promised that as we seek God and His Kingdom above all else, He will give us everything we need.[10]

These verses lead us to the next time we will actually see the Tree of Life. The redeemed will arrive in the heavenly city where the throne of God and the Lamb (Jesus) are. From it, a river of water of life flows down the center of the main street. "On each side of the river grew a tree of life, bearing twelve crops of fruit, with a fresh crop each month. The leaves [are] used for medicine to heal the nations."[1]

The nations are people from over the whole world who accept Christ as their Savior.[12] Washing their robes is a metaphor that illustrates the individual's acceptance of Jesus. It refers to people who "have washed their robes in the blood of the Lamb and made them white."[13] This is a symbol of being covered by the righteousness of Jesus and being made holy by God as well as return to the image of God.

From this point on, there would "no longer...be a curse on anything."[14]

THE TREE OF THE KNOWLEDGE OF GOOD AND EVIL

At the second tree in the middle of the garden[15] was the Tree of the Knowledge of Good and Evil. Eve was forced to grapple with a need she hadn't recognized as a part of her life. She never considered she needed something she didn't already have. The beast was ready to make his play for humanity.

God instructed, "Of every tree of the garden you may freely eat; but of the tree of the knowledge of good and evil you shall not eat, for in the day that you eat of it you shall surely die."[16] The devil insinuated this was not true.[17] He told her the tree would make her like God.[18] The knowledge it gave was not good for her to have. She was not supposed to know good and evil.[19] Now through mistrust and disobedience, mankind had joined Lucifer.

How would God deal with this new defection? "Rebellion was not to be overcome by force. The Lord's principles are not of this order. His authority rests upon goodness, mercy, and love; and the presentation of these principles is the means to be used. God's government is moral, and truth and love are to be the prevailing power."[20]

Why was this turmoil permitted to come to the utopian planet?

THE ARENA WE LIVE IN

As we have already noticed, an event of enormous consequences had occurred in another idyllic location—heaven. Lucifer accused God of not being just. Such an event had never happened before. The other inhabitants of the universe would each make their own determination of who was right. God is on trial across all the expanses of existence. Citizens of the universe must witness how this plays out. An arena was determined where the rebellious angel could prove what his form of leadership would accomplish.

The stadium had to be in fair and neutral territory. The players had to be nonpartisan. Their introduction to both the battling sides could not be affected by previous experience that might prejudice their decisions. There was too much at stake. Eternity and universal peace were on the line. Everything had to be carried out in a manner that no one could claim a rigged outcome. Such an assertion would mire all life in never-ending dissension. Earth was already the field of contention. It would become the front line. The war in Heaven was ushered to Earth.

God allowed Lucifer access to the Garden of Eden at one site. He could only attack at the Tree of Knowledge of Good and Evil. It was undoubtedly a beautiful place. Everything God made was perfect. But why was Lucifer sent to this location? Why wasn't he allowed to go everywhere? Why was he even given a presence in one place? God permitted the testing of Lucifer's claims of unfairness. If the challenge of sin could be seen everywhere on earth, there would not have been any unprejudiced testimony to the goodness of God.

God wanted the universe to see the actual result of Lucifer's dissension. In His wisdom, God allowed Lucifer to have a place where he could speak to man. Man would choose to listen to God or Lucifer. For their safety, Adam and Eve were told not to go to the tree.

THE MOTIVE ISN'T CLEAR

We don't know why Eve approached the tree, but on approach, she fell to the enticements of Lucifer. "The serpent said to the woman, 'You will not surely die.'"[21] Remember God warned Adam and Eve, "If you eat of the tree, on that day you will surely die."[22] The devil's lie about death is one he has told ever since that time. He continues to say to people through unwitting messengers that they will not die.

The woman lacked discernment. She was offered "open eyes" to "be like God, knowing both good and evil."[23] Satan tempted her with the same ambition he succumbed to—to be like God.[24] Eve interpreted this offer in her mind to say it would give her wisdom. She wanted that and saw that the fruit was good to eat.[25]

God told Eve one thing, but she, in her mind, decided something else was true. God told her not to eat the tree's fruit. She told herself, "That looks good. It certainly looks pleasant. And it will make me wise."[26] When we believe that we know better than God, it will always lead us to great difficulties.

Did Eve even know what it meant to be wise? The concept probably hadn't been discussed. Made in the image of God, she had to have been wise, at least in some areas. But the serpent was able to change God's words and confuse her. She had yet to learn the lesson that "people do not live by bread alone; rather, we live by every word that comes from the mouth of the Lord."[27] The words God uses are critical. His meaning must be rightly discerned.

She also gave to her husband with her, and he ate.[28] Life on Earth changed at that moment.

On September 11, 2001, as news outlets reported on two planes flying into the World Trade Center in New York City, many people said, "Today the world changed."

But that change wasn't nearly as significant or dramatic as what happened on the day Eve ate the fruit. The whole earth changed at that moment. The Creator loved Adam and Eve and spent time with them. They chose to listen to a voice they didn't know.

For us today, however, man no longer has direct access to the Tree of Life and all its specified benefits. It is because of man's actions at the second tree.

God's ability to look into the future meant He knew this would happen. To save the creatures He loved, God established the rescue plan long before Earth's creation. He deployed it in the future. He sent Jesus His Son to live and to die for us.

By the time Eve ate the fruit, Jesus had already paid the price in the future. God said that turning away from the Giver of life would automatically result in death. When we choose to believe evil instead of Love, we become evil.

Because Man sinned, God had to have already put His plan into effect. To save man, God put the curse that should have been on man, into the ground. Everything in it began to die.[29] But man's salvation was already secured.

THE THIRD TREE IS THE CROSS

The first two trees reveal the only two choices we have available. Of course, those choices come wrapped in myriads of different coverings. In the end, they are only these: obedience and Life, or disobedience and death. We have precisely the same options that our first parents enjoyed. They are just this simple. Then, Jesus changed everything.

Jesus paid the price for the sins of Adam and Eve, and for your sins and mine. He did it on a cross. This Cross is the third tree. After Jesus died, the grave claimed its prize of the One who lived without sinning, who became sin for us. After Jesus rose, death no longer held power over Him. He proved that He could thwart sin's curse. The curse was defeated.

"For the wages of sin is death, but the gift of God is eternal life in Christ Jesus our Lord."[30] Because of what Jesus did before the foundations of the earth, we have the ability to conquer sin. He covers us with His goodness. He covers us with His triumph. He covers us with His righteousness.

At some point, sin and Lucifer will be destroyed. Jesus paid the price and demonstrated His victory over sin. Therefore, He's exposed Lucifer's chosen form of government, preferred form of leadership, accepted form of livelihood, as an evil form of life. Jesus holds the authority to overcome, and to put all that away, to wipe the slate clean. A time will come when evil is vanquished once, and for all.[31] God, Himself will prevail over all of what has been condemning us to death since Creation.

People ask, "Why does God allow for this to go on? If He is powerful and loving, why does He let so many people suffer in so many ways." When God created this earth, He made the mechanism of time to limit the duration of sin's control in the world. If time didn't exist, God could not have constrained the perpetuation of sin.

God allowed sin to take its course as all heavenly beings watched the drama playing out in this theater of the universe. He had to show results from the trajectory of Lucifer's ideas to their end, the effects of Lucifer's lies, and the suffering and misery sin would cause.

The years since Creation have been only a short span of time in relation to eternity. When time on earth ends, those who have chosen to follow Jesus will spend the rest of eternity with God, not having to worry about sin anymore. "Now I saw a new heaven and a new earth, for the first heaven and the first earth had passed away. Also there was no more sea…And God will wipe away every tear from their eyes; there shall be no more death, nor sorrow, nor crying. There shall be no more pain, for the former things have passed away."[32]

Our journey began on Creation Day 1, when the Lamb, Jesus, was the Light. Now we have come to the end when the re-creation of Planet Earth takes place. Once again, the Lamb, Jesus, is the Light that lightens all of humanity. In that time, we will be with God, and He will be our Light forever.[33]

"The nations of those who are saved shall walk in its light, and the kings of the earth bring their glory and honor into it. Its gates shall not be shut at all by day (there shall be no night there). And they shall bring the glory and the honor of the nations into it. But there shall by no means enter it anything that defiles, or causes an abomination or a lie, but only those who are written in the Lamb's Book of Life."[34]

The culmination of the Operation Time Box story is the reason people invite others to accept Jesus and allow the re-creation of His image in them. Soon, Jesus will return to take His people to spend eternity with Him in Heaven and the new earth. Everyone who is there has chosen to be in Jesus' presence. Only those who allowed Jesus to make the necessary changes in their hearts, to be recreated in His image, will stand with Him. God is eager for us to let Him recreate our lives and characters. With anticipation, He waits for us to accept Him. Are you ready to ask Him, "Will you change me, so I will be ready for heaven?"

Chapter 15

SO, WHAT'S LEFT?

So-called "prophets of doom" have been warning of the end of the world for many years. They have been scoffed at for longer than I can remember. There is something different about what we are seeing now.

Overlapping warnings from science of our likely demise are crucial to pay attention to. We have discussed the present decline of humanity's ability to function, and thus to survive, due to genetic entropy. There is no way to stop this from happening. It will grow to debilitating levels if we are around long enough.[1] In the same time frame, artificial intelligence is increasing in strength. Some are warning that Artificial Intelligence (AI) will reach "singularity" by 2045.[2] It means that computers will be so much smarter than us that they may decide people are not needed in their new bot-world.

In a way, the fact that the warning isn't that it will happen in thirty years gives it more credence. The thirty-year prophecies mentioned in Chapter 5 used by the environment/anti-God crusaders are well-chosen. A person can see that the dreaded event could happen in their own time, or at least in their child's time. A sense of urgency draws their full attention to the suggested solutions of the crisis. This usually takes the form of giving money and changing their lives to "save the world." Man believes he can save the world.

This subtly takes saving the world out of God's hands. It gives the impression that God is no longer needed, if He ever was.

On the other hand, if the forecasted catastrophe doesn't happen, most people will forget about it having been promoted. It is well known that we have short memories when it comes to politics and last year's crisis. Besides, it will have been replaced by new warnings for different calamities. They tend to blend into one. People don't remember that the thirty- and forty- and fifty-year-ago events never came to fruition. The doom-mongers can safely promote their new tragedy. This is the work of false science. It gives true science a bad reputation.

The present warnings of very likely scenarios (extinction-by-mutation and bot-world) seem not to have been able to gain political or financial advantage. Because of this, very little that is said about them makes it to mainstream media. After all, it would be unfortunate if their disclosure caused a drop in financial support of their research. If there is nothing we can do about these end-of-the-world scenarios, why spend a lot of money and risk bringing it on earlier? Is that cynical?

MAN SEEKING TO CHANGE THE SCALE OF BEING

This should activate our awareness to other ways that man is working to remove God from our consciousness. Let's review some, and the ways they work to dislodge God from his position at the top of the universal scale of being.

PROTECTING THE INANIMATES

Water, air, soil, minerals, plants, and weather are placed in legal status that raises their care and preservation above the needs of man.

Environmentalism has its roots in something other than species survival and "protecting the wild places of the earth."[3] Environmental groups such as the Sierra Club and others "have colluded with federal agencies in 'sue and settle' lawsuits."[4] In many cases, very little justification is needed to sue the government to protect a natural resource. While some of them have accomplished beneficial preservation, it is often at the expense of people who were excluded from the resource.

PROTECTING ANIMATED LIFE

Insects, reptiles, fish, birds, small and large animals have been given protected status that ignores the rights of people for the sake of life-forms lower in rank on the scale of being.

A famous example of this involved the snail darter in 1973.[5] This is a three-inch fish that was discovered in the waters to be affected by the building of the Tellico Dam on the Little Tennessee River. The dam, which was 99 percent finished at a cost of over $100 million, was to provide 200 million kilowatt hours of hydroelectric power. Construction was slowed. The intention of the environmentalists was that it be shut down and destroyed.[6]

Senator Howard Baker made an impassioned appeal to the Senate. He said, "We who voted for the Endangered Species Act with the honest intentions of protecting such glories of nature as the wolf, the eagle, and other treasures have found that extremists with wholly different motives are using this noble act for meanly obstructive ends."[7]

His amendment passed. Snail darters were transplanted to other rivers, and the dam was opened. As of 2013, the snail darter remains a threatened species,[8] but it does remain.

WHAT WE CAN'T HELP

These efforts are said to save these animals from extinction. It is true that man has not been faithful to our commission to subdue and have dominion over[9] the earth and the animals in it. In many cases, we are guilty of neglecting this responsibility. However, we have learned that it is likely that many animal extinctions should be blamed on genetic mutations rather than by human behavior or negligence. Entropy cannot be stopped by man, no matter what we do.

We do need to be better at managing the environs and creatures of nature. We also need to protect the environmental movement from extremists who place nature unnecessarily before man in importance.

SAME-SEX MARRIAGE

Our western culture (as well as cultures of most developed and developing countries) mocks sexual morality in many ways. One powerful way is by use of the powerful entertainment industry. By example, it makes the anything-goes mentality acceptable. God and His followers are mocked at any suggestion that there may be legitimate reason to avoid illicit behavior and even the media that promotes it.

The U.S. Supreme Court made same-sex marriage legal in every state on June 26, 2015.[10] This is still a matter of controversy around the country. Some Christians and other conservatives question what this means for the future of marriage.

Lesbian activist and journalist Masha Gessen voiced her opinion on a live radio broadcast in 2012. "It's a no-brainer that the institution of marriage should not exist. ...(F)ighting for gay marriage generally involves lying about what we are going to do with marriage when we get there—because we lie that the institution of marriage is not going to change, and that is a lie."[11,12]

The majority of people in the LGBT+ community may have no malice toward the institution of marriage. It is clear that at least one of the more militaristic leaders of the movement has intentions to wipe out, or at least to change the institution.

ABORTION

Abortion removes the sanctity that Creation built into childbirth. The entire process from love and marriage to sexual relations to childbirth and the family and education are part of the procreative mandate given to mankind. The purpose of all these combined in a process was to carry on the image of the loving God throughout the world.

Abortion introduces skepticism as to the love and wisdom of God in His plan for continuity of man. By questioning whether any unwanted child should be brought into the world, it transfers guilt from sinful egg and sperm donors to God for creating such a plan. It also eliminates any concept of morality. All this begs the existence of God.

SCIENCE THAT DOESN'T CONSIDER ALL EVIDENCE

Science builds on the work of previous discoveries. This is only natural and acceptable. However, [scientific theories of the past] become obsolete at a predictable rate. Harvard mathematician Samuel Arbesman calls this the "half-life of facts."[13] Dr. Richard Horton, editor-in-chief of the *Lancet*—considered to be one of the most well-respected peer-reviewed medical journals in the world—signs in on this. He says, "The case against science is straightforward: much of the scientific literature, perhaps half, may simply be untrue…science has taken a turn towards darkness."[14]

Many of the theories on evolution from two hundred years ago and much more recent have been shown to be questionable at best and even wrong. Still, they continue to be used as assumptions in other research. If the foundations for new research is wrong, the results will be wrong.

Many scientific fields seem to be interested in proving only what will benefit their existence. Population geneticists refuse to be honest with themselves and the world about our origins and the evidence and implications of genetic entropy. The most likely reason is their unwillingness to recognize a Creator. In the case of genetics, they won't allow true scientific method to direct their own personal preconceptions about their primary axiom[15] of evolution and natural selection. They seem not to be scientists at all. These methods show them to be neo-Darwinian religionists using selective scientific method to direct and promote their faith.

WARNINGS FROM SCRIPTURE

All this should ring clarion warning alarms to the typical man on the street. Predictions of these very types of occurrences have been made at least for the past 2500 years. The writer of the Old Testament book of Daniel was told, "But you, Daniel, keep this prophecy a secret; seal up the book until the time of the end, when many will rush here and there, and knowledge will increase."[16]

Nearly five hundred years before that, Israel's King Solomon wrote, "The greater my wisdom, the greater my grief. To increase knowledge only increases sorrow."[17] These prophecies were not intended to prevent the increase of knowledge or to subvert it, but only to clue us in that at the time of the great increase in knowledge, the end will come.

A TIME OF JUDGMENT

We return to our introduction where a very important reason was given to study this material. The worship of God as Creator is among the most important issues shortly before Jesus returns. It is found in Revelation 14:7. An angel said with a loud voice, "Fear God and give him glory, because the hour of his judgment has come. Worship him who made the heavens, the earth, the sea and the springs of water."[18]

We do not worship Him in only one part of creation, but in all aspects of it. Some have singled out only one component as a representation of how to worship God. But every facet of the narrative of God as Creator is critical for us to focus on to worship Him fully. If not, the evil one would not be working so hard to destroy God's fingerprints in all of it.

Seven verses later, Jesus is coming. "Then I looked, and behold, a white cloud, and on the cloud sat One like the Son of Man, having on His head a golden crown, and in His hand a sharp sickle."[19] He is coming to bring in the harvest of souls who believe Him and accept His gift to us. The reminder is, the time is coming very soon. We will be judged to see who honors Him. Those who do will worship Him as Creator.

WORSHIP GOD AS CREATOR

Worship is the surrender of ourselves to the object of our worship. We are not the object of our worship. Neither are our music, the place we worship, the method of worship, the clothes we wear, the cars we drive, nor the order of our service, or anything else. Worship is all about focusing on God, surrendering to Him, and giving Him honor and praise for who He is.

We do not worship the things He made, or the abilities or talents He gave us. Our intelligence, the things we make with it, the theses we believe must not be worshipped. The most gifted among us has nothing on God. Neither will the bot that some fear will be smart enough to keep us from hitting the kill switch before it wipes us out.

Since Jesus was the active agent in creating us, we worship Him as Creator by accepting Him as our Savior and surrendering to Him. Surrender is giving up everything. It is becoming a slave to Jesus to be and to do what He asks of us for His good pleasure. Surrender is the basis of all worship, no matter which of the two great powers in the universe we choose to worship.

THE OTHER CHOICE

We either worship God or we worship the evil one. The lesser option is to the one who made nothing. The one who led angels to rebellion and got them expelled from Heaven. He also led mankind to rebel, getting us expelled from Eden. This resulted in hard labor for men and for women. It resulted in the mutation and deterioration of all life, and the eventual death of everything.

Had Jesus not been active in the affairs of man and earth since the beginning, everything would have been destroyed long ago. Still, some choose to worship one who only wishes to use us to hurt the One who loves us. Once sin is destroyed, those who want to continue in it would never be happy. For that person, Heaven would be a miserable place to spend eternity without the ability to be sinful. It would be a prison. God is not willing to punish anyone forever.

Some people don't like the idea of becoming a slave to anyone. We are already slaves to someone whether we recognize it or not. It is either God or the devil. This, again, is our choice. "Don't you know that when you offer yourselves to someone as obedient slaves, you are slaves of the one you obey—whether you are slaves to sin, which leads to death, or to obedience, which leads to righteousness?"[20]

God's safeguard to prevent the possibility of a recurrence of sin is to take only those who totally surrender and worship Him to join Him in His perfect eternity. As we worship Him, He is enabled to rebuild His image in us as it was intended to be in the beginning. Without the influence of evil, sin will not return.

WORSHIP AND CREATION

How does this apply to creation? If we worship Creator God, we commit ourselves to worship Him as He wishes to be worshipped. Worshiping God as Creator is putting all things into perspective. It is remembering that no matter what else there is, or how much we like it, we put God first. We do what He asks us to do. We don't take this action because we are afraid of Him. It is because we see His love for us and know that He is worthy to receive our adoration.

SCALE OF BEING

To worship God is to keep Him at the top of the Scale of Being. Nothing else and no one else will receive that place in our attentions. Nothing will be more important to us than Him. We will do all we can do, knowing that we can do nothing without Him giving us the will and the strength to do it. And that "[we] can do all things through Christ who strengthens [us]."[21]

He thinks so highly of us, His love is so great for us, that He made us in His own image. He wants all the other beings in the universe to see us and say, "There goes one who is in Creator's image. How God loves him/her!" Our joy will be to praise the One of whom we are merely an image. If we can live with that kind of love, eternity will see us living blessed lives. No matter what we must go through here, if we make the choice to worship Him as Creator, He will give us a place of honor in the universe and among the angels. How could we ask for more than that? How can we choose not to accept such a gift from such a God who loves us that much?

As we have noticed several times, we have been created in God's image. What does this have to do with worship? If we consider ourselves to be lower than what God esteems us to be, we are also diminishing (in our own eyes) the God in whose image we were created.

Part of the way we worship God as Creator is to accept the high rank in which He made us and live according to the value He placed on us. We choose to allow Him to build our character and our actions, to allow Him to recreate us in His image. To do less is to refuse to worship God as He deserves.

WILL THE WORLD BE DESTROYED?

Man is telling us through science the myriads of ways they believe humans may be destroyed. This is used to motivate us to comply with social and political agendas. Each of these agendas reduces our rights and abilities to manage our lives in a manner consistent with Heaven's eternal plan for us. Who wins with such a strategy? Folding back the layers of each such scheme will show a person or group of people who will receive the kickbacks of wealth and/or power.

Their warnings have included varieties of the flu, other viruses and bacterial infections, nuclear explosions, an ice age, earthquakes, hurricanes, tornadoes, wildfires, warming, cooling, drought, flooding, entropy, AI, and many others. There is reason to be concerned about some of these calamities. Certainly, there will be natural disasters. We see them regularly. There is no reason to believe they will cease to occur. However, none of these will destroy the world or mankind.

The world will be destroyed. It will not be by anything man has done or can do. God reserves for Himself the task of replacing what He has made as a part of Creation's rescue mission. The replacement will renew the "forever" promise.

An objection has long been made by scoffers saying, "This will never happen. It was written thousands of years ago and there is no evidence that it will ever come true." Part of the evidence is in the escalation of the number and frequency of the natural and man-made disasters Matthew 24 and 25 contain warnings of many events we are seeing in our time. The reason, however, for what seems like a delay in the fulfillment of the promise is given by Peter, the disciple of Jesus. "The Lord isn't really being slow about his promise, as some people think. No, he is being patient for your sake. He does not want anyone to be destroyed but wants everyone to repent."[22]

THE SCENARIO

At some time after the dispersal of the reminder/warning to worship God as Creator, Jesus will stand up from His seat at the throne of God and of the Lamb and say, "He who is unjust, let him be unjust still; he who is filthy, let him be filthy still; he who is righteous, let him be righteous still; he who is holy, let him be holy still."[23]

Then Jesus begins His harvest[24] only seven verses after the warning/reminder is given. All this shows us that the worship of Creator

God and Creation itself are key issues for us to focus on as we approach the harvest of souls.

Jesus' coming to rescue all the people who believed in Him is described in 1 Thessalonians. "For the Lord Himself will descend from heaven with a shout, with the voice of an archangel, and with the trumpet of God. And the dead in Christ will rise first. Then we who are alive and remain shall be caught up together with them in the clouds to meet the Lord in the air. And thus we shall always be with the Lord."[25]

The Book of Daniel gave more information about the rescue mission. "Then there will be a time of anguish greater than any since nations first came into existence. But at that time every one of your people whose name is written in the book will be rescued. Many of those whose bodies lie dead and buried will rise up, some to everlasting life and some to shame and everlasting disgrace."[26]

In Brazzaville, when my young family and I witnessed the riot that broke out in front of our house, we faced imminent danger. We had no part in the dispute, but we were in the middle of the enmity and fighting.

I'm not sure our children understood the danger we were in. As the minutes ticked by that afternoon, we could have allowed our emotions to spiral out of control. We feared for the lives of our children and for each other. The wait seemed interminable, and our relief was overwhelming when the Marines arrived to take us out of the danger. The knowledge that a rescue mission was underway and that our God oversaw every event in our lives gave peace in the middle of an unnerving event.

This rescue is exactly what the world, led by the evil one, seeks to keep from happening. The one who brought sin into the world is responsible for the suffering of every person who has lived. He uses us as "collateral damage" to hurt God, who loves us with an everlasting love.[27]

His aim is to turn everyone away from God by whatever means possible, hoping that we will worship him instead. He gives us the same offer he made to Jesus in the wilderness of temptation. "Worship me, and I will not only give you the desires of your heart, but I will do it in an easier way than the way God has designed."[28]

A myriad of attacks in our day chip away at the image of God in us. This is not simply for the purpose of destroying a plausible explanation of our origins. It is to destroy the concept of God and erase any expectation of a standard morality. Organizations and individuals understand the importance of His image. It is why they want to destroy it.

It is my prayer that this study of Creation will draw our attention to God as Creator and will build our appreciation for Him and His glory.[29] It will teach us the character and other invisible attributes of God.[30] And by seeing that we are made in His own image,[31] we will understand the character that He wants to recreate in us.[32] God will do the work in us that we cannot do for ourselves.[33] When we surrender to Him, He will totally re-image us for eternity.[34] As long as time endures, all of Creation calls out to each human heart: "Worship Him who created!"

What will you decide? God is specifically and intentionally inviting you to worship Him. Why? Is it because He is a megalomaniac? Is it because He wants to have all the attention on Himself? Does it make Him feel good?

No. It is because He knows that if our attention is on Him, we will not be worshipping what brings only misery and death. That's why He says, "Choose today: life or death."[35] It is your choice. It is mine. That choice is portrayed in everything we do. That choice designates whether we will enjoy eternity.

DESTRUCTION OF ALL EVIL

At the time of Jesus' rescue mission, Daniel mentions some who will rise up to shame and everlasting disgrace. Their names are not written in the Book of Life that belongs to the Lamb (Jesus).[36] They will not be ready to be taken to Heaven. They are described as "men of lawlessness [who] the Lord Jesus will slay…with the breath of his mouth and destroy him by the splendor of his coming."[37]

Then, one thousand years after Jesus' rescue mission to Earth, a second resurrection will bring back to life all the wicked who ever lived. At that time, Satan will be released from the prison where he had been confined by a chain of circumstances since the time of the rescue.[38]

He will gather all his resurrected malcontents from all over the world and all time[39] and will seek to destroy God and His people.[40] They will be in the New City Jerusalem coming down from God out of Heaven.[41] The devil and His masses will not be able to defeat God. Fire will come from heaven and destroy them once and for all.[42]

This will create a lake of fire.[43] Into this fire will also be thrown death and hell and everyone whose name was not written in the Book of Life.[44] The righteous who were raised in the first resurrection and who were taken to Heaven by Jesus because their names are in the Book of Life will not be put into this fire.

THE END, THE BEGINNING

This fire will grow to consume the entire earth and every remaining vestige of sinful man. "The heavens will pass away with a great noise, and the elements will melt with fervent heat; both the earth and the works that are in it will be burned up."[45] This will destroy everything that sinful man ever did and made.

There will only be one place where a reminder of man's sin can be found. It is the scars in the hands of Jesus. If sin should ever begin to build up in the heart of any heavenly being, the indelible marks serve as an eternal reminder of the results of sin. The knowing universe will then have no question about whether God will be justified in eliminating any sin that may re-ignite the pain of all the years of man on Earth.

This was promised before Jesus came. "What do you conspire against the Lord? He will make an utter end of it. Affliction will not rise up a second time."[46]

But the destruction of the earth is not what we are looking for. "We are looking forward to the new heavens and new earth He has promised, a world filled with God's righteousness."[47] The beauty and glory will be as it was on Day 7. There will be some major differences.

From Day 1. The elements of the land and the water will be destroyed by the consuming fire.

From Day 2. Likewise, the elements of the atmosphere will be gone.

From Day 3. There will be no seeds in the plants. They will not be needed. Without sin, plants won't die, so the newly created replacement plants will not have them.

From Day 4. "There shall be no night there: [His servants] need no lamp nor light of the sun, for the Lord God gives them light. And they shall reign forever and ever."[48]

From Day 5. Animals will not reproduce. With no death, the animals God creates in Heaven and the New Earth will live forever.

From Day 6. Man will not reproduce. There will also be no need to have the institution of marriage to point to God's relationship with man. We will dwell with God and enjoy that relationship face-to-face. "When the dead rise, they will neither marry nor be given in marriage. In this respect they will be like the angels in heaven."[49] The prime purposes for marriage will be fulfilled. This makes marriage and the marital relationship types of our worship relationship with God.

There will be no death. As with the animals, there will be no need to reproduce. We will not need birth to point us to the promise of the coming of Jesus. We will be in His presence. I have heard some say, "If there is no sex in Heaven, I don't want to go." We should rest assured, that God will have a replacement that will be even more gratifying and fulfilling than sex.

After all, real love is not the feeling that comes from our experiences. Everything that was made during Creation points us to the God of love. It is all part of our worship of Him. In Heaven, our worship will be undefiled by sin in our hearts or by the separation from Him we live with now. It will be direct. Face-to-face. Rest assured that worshipping Him will be far more gratifying than any sexual experience.

WHAT WILL REMAIN?

On Day 7, the Sabbath was made by making a day holy. The Bible tells us we will continue to worship God on the Sabbath. "'For as the new heavens and the new earth which I will make shall remain before Me,' says the Lord, 'So shall your descendants and your name remain. And it shall come to pass that from one New Moon to another, and from one Sabbath to another, all flesh shall come to worship before Me,' says the Lord."[50]

ONE MORE THING

The fact that the Sabbath remains is a bit perplexing. That Time Box created "in the beginning" was made to contain sin and its originator. We see that sin and the evil one will be destroyed by the fire that melts the elements of earth before it is re-created. We will no longer have the sun to give us light. It is what defined our days.

Before the coming of Jesus, in Revelation 10, an angel was seen "stand(ing) upon the sea and upon the earth lifted up his hand to heaven…" He brought attention back to Creation as if to remind us that the end has everything to do with the beginning. "And swore an oath by him that lives for ever and ever, who created heaven, and the things that are in it, and the earth, and the things that are in it, and the sea, and the things which are in it, that there should be time no longer."[51]

From all Creation, there are only two things that will be seen in Heaven and the New Earth. Only two things that are so valuable to God that He carries them from time to eternity. They are Mankind and the Sabbath. The Sabbath changes from being a portal from time to eternity to being a direct portal to God for eternity.

Then, there is You. This whole rescue mission focuses on You. Jesus said, "Let not your heart be troubled; you believe in God, believe also in Me. In My Father's house are many mansions; if it were not so, I would have told you. I go to prepare a place for you. And if I go and prepare a place for you, I will come again and receive you to Myself; that where I am, there you may be also."[52]

When it is all said and done, it's all up to You.

Are You in?

EPILOGUE

So, from a single, slatted porch in the Adirondacks—or nearly any other mountain range in the world—the seven-year-old and the biophysicist sit together in rocking chairs. Throughout the year they view amazing splendor and diversity in the exact same scene.

At one time, it is the sparkle on the lake waters' surface with dramatic white clouds in dazzling blue skies over luscious green slopes of pine and maple trees, topped with the jagged rocky cliffs above.

Just a few weeks later, the same scene displays the most magnificent reds and yellows and pastel greens of a Master Painter's palette.

Then, from the same easy chairs, the beautiful "white Christmas" glory of snow blankets everything in a somehow warming chill that prepares the scene to start again.

And every night the stars and planets dance with the moon from horizon to horizon to display the temptation to consider something bigger than what the eye can see.

NOTES

INTRODUCTION
1. Genesis 1:1 NKJV
2. Romans 1:20 NIV
3. Genesis 1:26 NIV
4. Revelation 14:7 NIV

CHAPTER 1 CREATION, DAY 1 SAVE THE WORLD BEFORE YOU BUILD IT
1. Genesis 1:1 NKJV
2. Isaiah 66:23 NKJV
3. Revelation 21:23 NIV
4. Revelation 22:5 NIV
5. Revelation 10:5, 6 NKJV
6. Hebrews 11:3 KJV
7. https://biologos.org/common-questions/scientific-evidence/evolution-and-the-second-law/ Captured 2/7/18.
8. https://www.space.com/25100-multiverse-cosmic-inflation-gravitational-waves.html Captured 2/9/18.
9. Ibid.
10. Psalm 147:4 NKJV
11. Job 9:9, 38:32, 26:13, 38:31; Amos 5:8, Acts 28:11
12. Job 38:8
13. Genesis 1:2 NIV
14. Ibid.
15. https://www.space.com/14719-spacekids-temperature-outer-space.html Captured 3/13/18.
16. https://www.universetoday.com/112805/why-is-everything-spherical/ Captured 3/5/18.
17. Job 38:28-30 NIV
18. Genesis 1:2 NIV
19. Genesis 1:3-5 NKJV
20. John 1:1-5 (emphasis supplied)
21. John 1:14
22. John 9:5 NKJV "As long as I am in the world, I am the light of the world."
23. Revelation 21:22-25 NIV
24. Malachi 4:2
25. Job 37:10 NKJV
26. Psalm 147:18 NKJV
27. 2 Thessalonians 2:8 NKJV
28. John 9:5 NKJV

29. John 14:17 NJKV
30. Philippians 2:15, 16 NIV
31. Matthew 5:14 NKJV
32. Ephesians 1:4, 5

CHAPTER 2 CREATION, DAY 2 SETTING THE FOUNDATIONS IN ORDER

1. Genesis 1:5, 8
2. https://spaceplace.nasa.gov/days/en/, https://nssdc.gsfc.nasa.gov/planetary/factsheet/ Captured 3/13/18.
3. http://www.christiananswers.net/q-aig/aig-c003.html Captured 3/13/18
4. Ibid.
5. 1 Peter 2:9 NIV
6. The Trinity of God consists of God the Father, God the Son (Jesus), and God the Holy Ghost (or Holy Spirit). They can be found in Matthew 28:19, "Go ye therefore, and teach all nations, baptizing them in the name of the Father, and of the Son, and of the Holy Ghost."
7. Genesis 1:6-8
8. https://www.chemistryworld.com/news/space-ice-goes-against-the-grain/3003728.article Captured 2/12/18.
9. https://www.space.com/24292-rosetta-spacecraft.html Captured 2/12/18.
10. Ibid.
10 a Ibid.
11. https://www.space.com/33011-life-building-blocks-found-around-comet.html Captured 2/12/18.
12. http://earthsky.org/earth/what-is-the-source-of-the-heat-in-the-earths-interior (emphasis supplied) Captured 2/12/18.
13. 2 Peter 3:10 NKJV (emphasis supplied)
14. 2 Thessalonians 2:8
15. Ibid.
16. Isaiah 25:9
17. Revelation 20:9
18. Genesis 2:5 NKJV
19. Proverbs 8:27-29 NIV
20. Job 38:8-10
21. http://earthsky.org/earth/what-keeps-earths-atmosphere-on-earth Captured 2/12/18
22. Job 37:10, 11 NKJV
23. Jeremiah 10:12, 13
24. 1 Corinthians 14:40
25. Psalm 19:1 NKJV
26. Psalm 97:6 NKJV
27. Elwell, W. A., & Beitzel, B. J. (1988). *Righteousness. Baker Encyclopedia of the Bible* (Vol. 2, pp. 1860–1862). Grand Rapids, MI: Baker Book House. (LOGOS)
28. Romans 1:20
29. Romans 1:21
30. Eugene Linden, Time July 13, 1992, 62-66
31. Romans 1:25 NKJV
32. Revelation 14:7 NKJV

CHAPTER 3   CREATION DAY 2, PART B   GIVE ME ANOTHER GLASS OF THAT WATER
1.   Genesis 1:6, 7 NKJV
2.   Genesis 7:11, 12
3.   https://answersingenesis.org/bible-timeline/timeline-for-the-flood/ Captured 2/13/18
4.   Genesis 7:11 NIV
5.   Genesis 7:11, 12
6.   Genesis 7:18, 19 NIV
7.   http://timesofindia.indiatimes.com/home/science/Massive-underground-reservoir-of-water-three-times-the-size-of-Earths-oceans-located/articleshow/36493512.cms Captured 6/13/14.
8.   Ibid.
9.   https://www.newscientist.com/article/dn25723-massive-ocean-discovered-towards-earths-core#.U5skchZgvwI Captured 6/13/14.
10. http://timesofindia.indiatimes.com/home/science/Massive-underground-reservoir-of-water-three-times-the-size-of-Earths-oceans-located/articleshow/36493512.cms Captured 6/13/14.
11. Ibid.
12. Ibid.
13. https://phys.org/news/2014-06-evidence-oceans-deep-earth.html#jCp Captured 6/12/14.
14. https://earthobservatory.nasa.gov/IOTD/view.php?id=3499   Captured 2/13/18.
15. http://rt.com/news/165756-underground-ocean-discovered-water/ Captured 7/15/15.
16. Ibid.
17. Psalm 139:14 NKJV
18. Jeremiah 10:7
19. Jeremiah 10:12
20. Deuteronomy 4:28
21. 2 Peter 3:3-5
22. John 7:37 NKJV
23. John 4:14
24. John 19:28
25. Isaiah 53:2
26. Philippians 1:6 NKJV "Being confident of this very thing, that He who has begun a good work in you will complete it until the day of Jesus Christ."
27. Philippians 2:13
28. Philippians 1:6

CHAPTER 4 CREATION, DAY 3   WHAT A DIFFERENCE A DAY MAKES
1.   Genesis 1:9, 10
2.   1 Corinthians 3:11
3.   John 4:14
4.   Genesis 1:11-13 NIV
5.   Psalm 33:9
6.   Genesis 1:9 NKJV

7. Genesis 1:11 NKJV
8. Genesis 2:9 NKJV
9. http://www.colour-affects.co.uk/psychological-properties-of-colours Captured 2/14/18.
10. http://www.wisegeek.org/what-are-the-most-relaxing-colors.htm Captured 2/14/18.
11. Ibid.
12. Ibid.
13. http://www.huffingtonpost.com/2011/11/27/how-color-affects-our-moo_n_1114790.html Captured 2/14/18
14. White, E.G. *Thoughts From the Mount of Blessings*, Nampa, ID: Pacific Press Publishing Association, 2016, 96.
15. Genesis 1:11, 12
16. Genesis 1:29, 30
17. http://grisda.org/resources1/faq/age-of-the-earth/ "The chronological figures related to genealogies in Scripture add up to approximately 6,000 years since the creation described in Genesis. Many creationists consider these figures to be relatively complete, and thus the Earth is considered to be about 6,000 to perhaps as much as 10,000 years old."
18. Matthew 6:33 NIV
19. Revelation 22:2 NKJV
20. Strong, James A., *The Exhaustive Concordance of the Bible*, Worldwide Publishers, Iowa Falls, Iowa
21. Galatians 3:13 NKJV
22. Proverbs 14:12 NIV
23. Matthew 25:14-30
24. Luke 6:46-49

CHAPTER 5   CREATION, DAY 4   NO DARK VALLEYS

1. The Khmer Rouge were the adherents to the Communist Party of Kampuchea in Cambodia. Their leader was Pol Pot. Encyclopedia Britannica. Khmer Rouge. https://www.britannica.com/topic/Khmer-Rouge.
2. Joel 2:31 NKJV
3. Genesis 1:15-19 NKJV
4. 2 Thessalonians 2:10, 11, NKJV
5. 2 Thessalonians 2:12 NKJV
6. Psalm 105:28 NKJV
7. Exodus 14:19, 20
8. Genesis 1:18 NKJV
9. Jeremiah 33:19-21
10. Revelation 21:23 NKJV
11. Isaiah 60:19, 20
12. *Time*, June 24, 1974
13. *Time*, April 28, 2008
14. http://www.climatedepot.com/2018/02/20/noaa-caught-adjusting-usas-big-freeze-out-of-existence-fiddling-with-the-raw-temperature-data/ Captured 3/14/18
15. Ibid.

16. Deuteronomy 4:19
17. Revelation 14:7

CHAPTER 6 CREATION, DAY 5 THE SCALE OF BEING

1. White, E.G. *Desire of Ages*, Nampa, ID: Pacific Press Publishing Association, 2006, 313.
2. Genesis 1:20-23 NKJV
3. Genesis 1:1 NKJV
4. Genesis 1:3, 4 NKJV
5. Genesis 1:6 NKJV
6. Genesis 1:9 NKJV
7. Genesis 1:14 NKJV
8. Genesis 1:20 NKJV
9. Genesis 1:21 NKJV
10. Genesis 1:24 NKJV
11. Genesis 1:22 NKJV
12. Henry, M. (1994). *Matthew Henry's Commentary on the Whole Bible: Complete and Unabridged in One Volume*, p. 16. Peabody: Hendrickson. (Logos)
13. https://www.guttmacher.org/fact-sheet/induced-abortion-united-states Captured 3/14/18.
14. White, E.G. *Patriarchs and Prophets*, Nampa, ID: Pacific Press Publishing Association, 2002, 84.
15. Matthew 10:29-31
16. Revelation 14:7
17. Luke 12:22-31
18. Luke 12:32
19. Luke 12:33
20. Matthew 6:33
21. Luke 12:34

CHAPTER 7 CREATION, DAY 6A THE HIGHEST ORDER

1. Genesis 1:24, 25 NKJV
2. Genesis 1:26 NKJV
3. http://science.sciencemag.org/content/299/5612/1523.full Captured 3/14/18
   Science 07 Mar 2003:Vol. 299, Issue 5612, pp. 1523-1524 Captured 7/21/17
   Is evolution a religion as opposed to science? "A major complaint of the Creationists, those who are committed to a Genesis-based story of origins, is that evolution—and Darwinism in particular—is more than just a scientific theory. They object that too often evolution operates as a kind of secular religion...if we wish to deny that evolution is more than just a scientific theory, the Creationists do have a point." The article concedes the question on some points but misses the fact that evolutionary species development is still unsubstantiated wishing (faith) and not proven by scientific method.
4. http://evolution.berkeley.edu/evolibrary/article/lines_03 Captured 3/14/18.
5. http://rsos.royalsocietypublishing.org/content/3/8/160328 Captured 7/31/17.
6. https://www.theguardian.com/science/2009/oct/21/fossil-ida-nature-magazine-revelation Captured 7/31/17.

7. http://discovermagazine.com/2003/sep/cover Captured, 8/1/17.
8. Genesis 1:24 NIV
9. Genesis 1:26 NIV
10. Genesis 5:1-3 NIV
11. Genesis 1:27
12. See Isaiah 14:13, 14
13. Exodus 8:10
14. 1 Timothy 6:15, 16 NIV (emphasis supplied)
15. See Genesis 5:1-19
16. Genesis 3:19 NIV
17. Genesis 5:24 NKJV
18. 2 Kings 2:11 NKJV
19. Ezekiel 18:4 NKJV
20. 1 Thessalonians 4:14-17 NIV
21. Technically, people who lived before the time of Jesus did not believe in Him by name. However, they believed in the promise made by God in Genesis 3:15 that a Savior would come. Their belief was in Creator God, who we have asserted in Chapter 1 was God the Son, or Jesus, according to John 1:1-3, 15.
22. John 11:1-44
23. 2 Thessalonians 2:8
24. 1 Corinthians 15:52, 53 NIV
25. Philippians 3:20-21 NIV
26. 2 Peter 3:13
27. Revelation 21:1
28. Revelation 22:14
29. Genesis 3:22
30. Isaiah 66:2
31. Genesis 2:16, 17
32. Genesis 3:5
33. Genesis 3:22
34. Genesis 3:23
35. Psalm 139:7-10
36. Philippians 2:5-7 NIV
37. Philippians 3:21
38. Isaiah 14:14
39. Proverbs 16:18
40. Hebrews 13:5

CHAPTER 8   CREATION, DAY 6B   MANKIND IS IN CONTROL

1. Romans 10:9, 10
2. Romans 8:29 (emphasis supplied)
3. Romans 8:28 (emphasis supplied)
4. Matthew 18:14 NIV
5. Matthew 28:19, 20
6. Revelation 14:7 NKJV
7. https://www2.palomar.edu/anthro/evolve/evolve_1.htm Captured 2/22/18.

8. Ibid.
9. Ibid.
10. https://evolution.berkeley.edu/evolibrary/article/history_09 Captured 2/22/18.
11. http://sciencenetlinks.com/lessons/the-history-of-evolutionary-theory/ Captured 2/22/18.
12. https://en.wikipedia.org/wiki/Relationship_between_religion_and_science Captured 2/22/18.
13. https://www.biography.com/people/ben-stein-9542432 Captured 2/23/18.
14. http://www.intelligentdesign.org/whatisid.php Captured 2/23/28.
15. *Expelled: No Intelligence Allowed*, Ben Stein, Premise Media Corp., 2008.
16. *Is Genesis History?* Dell Tackett, Compass Cinema, 2017.
17. http://blog.godreports.com/2017/07/discovery-of-soft-tissue-in-dinosaur-bones-upends-evolutionary-timetable/ Captured 2/22/18.
18. http://blog.godreports.com/2017/08/university-settles-lawsuit-with-scientist-fired-after-he-found-soft-tissue-in-dinosaur-bones/ Captured 2/22/18.
19. Genesis 1:26 NIV
20. Genesis 5:3 NIV (emphasis supplied)
21. Genesis 1:26
22. https://www.nytimes.com/2014/04/27/magazine/the-rights-of-man-and-beast.html?mcubz=0 Captured 2/23/18https://www.nytimes.com/2014/04/27/magazine/the-rights-of-man-and-beast.html?mcubz=0 Captured 2/22/18.
23. https://www.theverge.com/2017/1/19/14322334/robot-electronic-persons-eu-report-liability-civil-suits Captured 2/13/18.
24. Ibid.
25. http://www.express.co.uk/news/science/871886/saudi-arabia-robot-sophia-artificial-intelligence-ai-citizenship Captured 11/3/17.

CHAPTER 9   CREATION 6C   SPECIAL KIND OF CREATION

1. Genesis 1:3, 6, 9, 11
2. Genesis 1:26 (emphasis supplied)
3. Genesis 2:19 NIV
4. Genesis 2:7 (emphasis supplied)
5. John 3:16, 17
6. Genesis 1:27
7. Genesis 2:18
8. Genesis 2:21, 22
9. White, E.G. *Desire of Ages*, Nampa, ID: Pacific Press Publishing Association, 2006, 313.
10. B'rachot 61a; Rashi to Genesis 2:22. Kaufman, Michael. *Love, Marriage, and Family in Jewish Law and Tradition.* Jason Aronson Publishers, 177.
11. Genesis 1:28
12. Matthew 28:19, 20
13. Romans 3:23 NIV
14. Romans 3:10
15. Romans 6:23 NIV (emphasis supplied)
16. Romans 5:8

17. Romans 3:24
18. https://en.wikipedia.org/wiki/Sexual_minority, Captured 9/26/17.
19. https://en.wikipedia.org/wiki/LGBT_demographics_of_the_United_States Captured 3/14/18.
20. Ephesians 5:25, 26
21. Ephesians 5:27 NLT
22. Sotah 17a. Kaufman, Michael. *Love, Marriage, and Family in Jewish Law and Tradition.* Jason Aronson Publishers, 178, 179.
23. 1 John 1:9
24. John 8:11 NIV
25. Genesis 1:28 NIV
26. Proverbs 22:6
27. Matthew 1:23
28. Galatians 4:4
29. Hebrews 4:15 NIV
30. John 3:16
31. Romans 1:20
32. Revelation 14:7
33. http://www.washingtontimes.com/news/2017/feb/5/climate-change-whistleblower-alleges-noaa-manipula/ Captured 2/26/18.
34. https://www.usatoday.com/story/news/politics/2016/07/14/republicans-say-abortion-clinics-broke-law-selling-fetal-organs/87078180/ Captured 3/14/18.
35. Jeremiah 31:3

CHAPTER 10   CREATION, DAY 7   PORTAL TO ETERNITY

1. Mark 2:27 NKJV
2. Genesis 2:3 NIV
3. Anil Kanda, Message at ACF Institute, May 27, 2017.
4. Genesis 2:1-3
5. Isaiah 58:13 NKJV
6. Leviticus 23:32
7. John 4:23, 24 NKJV
8. 1 John 1:9
9. 1 Corinthians 15:53 brackets supplied
10. Philippians 3:20, 21 NIV
11. Isaiah 58:14
12. http://quotes.yourdictionary.com/articles/who-said-absence-makes-the-heart-grow-fonder.html Captured 3/14/18.
13. Mark 2:27 NKJV
14. Genesis 3:15
15. John 14:3

CHAPTER 11   CREATION, DAY 7B   CREATING THE PORTAL

1. Genesis 2:1-3
2. Genesis 2:1 NKJV
3. Mark 2:27 NKJV

4. Revelation 21:1
5. Revelation 21:4, 5
6. White, E.G. *The Faith I Live By*, Hagerstown, MD: Review and Herald Publishing Association. 1999, 32, Paraphrased.
7. Genesis 1:22 NKJV
8. Genesis 1:28 NIV
9. Matthew 28:19, 20
10. Genesis 2:3 NKJV
11. Faithlife Corporation. (2018). to sanctify (Version 7.13) [Computer software]. Logos Bible Software Bible Sense Lexicon. Bellingham, WA: Faithlife Corporation. Retrieved from https://ref.ly/logos4/Senses?KeyId=ws.sanctify.v.01
12. Genesis 3:21 NKJV
13. Hebrews 10:10
14. 1 John 1:9 NKJV
15. Romans 8:38, 39
16. Revelation 22:11
17. Exodus 12:25-27
18. Genesis 2:3 NKJV
19. Genesis 1:28 NIV
20. Exodus 16:15 NKJV
21. Exodus 16:18
22. Exodus 16:27, 28
23. Exodus 20:8 NKJV
24. Exodus 20:9, 10
25. Exodus 20:10, 11
26. Revelation 14:7
27. Hebrews 4:4, 6, 9 NIV emphasis supplied
28. 1 Peter 4:7-9

CHAPTER 12   CREATION   AFTER THE FALL

1. Richard O'Ffill was the Director of the Seventh-day Adventist World Service (S.A.W.S.), now known as the Adventist Development and Relief Agency. ADRA delivers relief and development assistance to individuals in more than 130 countries—regardless of their ethnicity, political affiliation, or religious association. https://adra.org/about-adra/
2. Isaiah 14:12
3. Isaiah 14:13-15
4. Ezekiel 28:12-15
5. Revelation 12:7, 8 NIV
6. Revelation 12:9 NIV
7. Revelation 12:4 NIV
8. Revelation 13:8
9. Genesis 2:15-17 NKJV
10. Genesis 6-9
11. Exodus 12:1-30
12. Matthew 13:24-30

13. Matthew 25:32, 33
14. Revelation 7:2, 3 NJKV
15. Revelation 13:17; 14:9-11; 15:2; 16:2 NKJV
16. Genesis 3:4, 5 NKJV
17. Genesis 2:16, 17
18. Genesis 3:6, 7
19. Genesis 3:14
20. Genesis 3:15
21. Matthew 1:23 NKJV
22. Genesis 3:7-10

CHAPTER 13 CREATION WHAT'S A SAVIOR TO DO?

1. Genesis 3:16, NIV
2. Hyatt, Michael. *Your Best Year Ever*, Baker Books, 2018, 31.
3. Genesis 3:17-19a NKJV
4. Matthew 11:28, 29 NKJV
5. Genesis 3:18,19
6. Genesis 2:2
7. Exodus 20:8, 9 NIV
8. Galatians 3:13
9. Deuteronomy 21:22, 23
10. 2 Corinthians 5:21
11. Galatians 3:13 NKJV
12. Romans 5:12
13. 1 Corinthians 15:56, 57
14. Revelation 13:8 NIV
15. Genesis 2:17 NKJV
16. Luke 24:5-7 NIV
17. Hosea 13:14 NKJV
18. Genesis 3:15 NIV
19. Romans 8:20-22
20. Genesis 1:29 This diet was altered after sin (Genesis 3:18), and again after the Flood (Genesis 7:2, 8)
21. Genesis 3:17-19
22. Genesis 2:7 "Then the Lord God formed the man from the dust of the ground. He breathed the breath of life into the man's nostrils, and the man became a living person."
23. Sanford, https://www.youtube.com/watch?v=K8KbM-xkfVk @ 7:23
24. Sanford, https://www.youtube.com/watch?v=pJ-4umGkgos @ 9:17
25. Sanford, https://www.youtube.com/watch?v=K8KbM-xkfVk @ 6:29
26. Sanford, Ibid. @ 5:55
27. Sanford, Ibid @ 7:07
28. Michael Lynch, *Genetics*, March 1, 2016 vol. 202 no. 3 869-875; https://doi.org/10.1534/genetics.115.180471
29. Ibid.
30. Ibid.

31. Ibid.
32. Ibid.
33. Institute of Evolutionary Biology, School of Biological Sciences, University of Edinburgh, Edinburgh, United Kingdom
34. http://onlinelibrary.wiley.com/doi/10.1111/evo.12195/full
35. Sanford, https://www.youtube.com/watch?v=K8KbM-xkfVk @ 2:56
36. Sanford, *Genetic Entropy*, 2014 p. 211 on Lynch 2010.
37. Romans 6:7
38. Genesis 1:26 NIV
39. 2 Corinthians 5:17 NKJV
40. 1 Corinthians 15:53
41. White, E.G. *Patriarchs and Prophets*, Nampa, ID: Pacific Press Publishing Association. 2002, 82.
42. 1 Corinthians 15:24-26 NIV
43. Revelation 20:9,10,14 NKJV
44. What is the second death? http://drgarywagner.com/resources/what-is-the-second-death/
45. Isaiah 25:8 JKV
46. 1 Corinthians 15:54 NLT
47. Revelation 21:4
48. What is the meaning of eternal punishment, unquenchable fire, tormented day and night, forever and ever? http://drgarywagner.com/resources/what-is-the-second-death/

CHAPTER 14   CREATION–TWO GARDENS AND THREE TREES

1. Genesis 1:28 NKJV
2. Rev 21:4 NKJV
3. Genesis 2:9 NKJV
4. Genesis 3:22
5. Proverbs 3:18, 19
6. Proverbs 11:30
7. Proverbs 13:12
8. Proverbs 15:4
9. James 1:5 NKJV
10. Matthew 6:33
11. Revelation 22:1, 2
12. Revelation 22:14
13. Revelation 7:14
14. Revelation 22:3
15. Genesis 2:9 NKJV
16. Genesis 2:17 NKJV
17. Genesis 3:4 NKJV
18. Genesis 3:5 NKJV
19. Genesis 3:22 NKJV
20. White, E.G. *Desire of Ages*, Nampa, ID: Pacific Press Publishing Association, 2006, 759.

21. Genesis 3:4 NKJV
22. Genesis 2:17 (paraphrased)
23. Genesis 3:5
24. Isaiah 14:14 NKJV
25. Genesis 3:6 NKJV
26. Genesis 3:6 (paraphrased)
27. Deuteronomy 8:3 Matthew 4:4 (paraphrased)
28. Genesis 3:6 (paraphrased)
29. Genesis 3:17, 18 (context)
30. Romans 6:23 NKJV
31. See Revelation 20:9, 10, 14 NKJV
32. Revelation 21:1,4 NKJV
33. Revelation 21:22, 23 NKJV
34. Revelation 21:24-27 NKJV

CHAPTER 15 CREATION SO, WHAT'S LEFT?

1. My belief is that Jesus will return before, or at the time the Singularity occurs. He will not allow mankind to be destroyed by any power.
2. Called the Singularity. http://metro.co.uk/2017/07/27/the-end-of-humanity-as-we-know-it-is-coming-in-2045-and-big-companies-are-working-towards-it-6807683/ Captured 2/1/18.
3. Part of the Sierra Club's mission as stated on their IRS Form 990 in 2015. Downloaded from https://www.activistfacts.com/organizations/194-sierra-club/ 2/1/2018.
4. Ibid.
5. https://en.wikipedia.org/wiki/Snail_darter_controversy Captured 2/7/18.
6. Ibid.
7. Ibid.
8. https://ecos.fws.gov/docs/five_year_review/doc4136.pdf Captured 2/7/18.
9. Genesis 1:28
10. https://www.nytimes.com/2015/06/27/us/supreme-court-same-sex-marriage.html Captured 2/7/18.
11. https://illinoisfamily.org/homosexuality/homosexual-activist-admits-true-purpose-of-battle-is-to-destroy-marriage/ Captured 2/7/18.
12. http://www.abc.net.au/radionational/programs/lifematters/why-get-married/4058506 Captured 2/7/18.
13. http://qi.com/research-half-life-of-facts Captured 2/7/18.
14. http://www.thelancet.com/pdfs/journals/lancet/PIIS0140-6736%2815%2960696-1.pdf April 11, 2015 Captured 2/7/18.
15. Sanford, *Genetic Entropy*, p. v.
16. Daniel 12:4 emphasis supplied
17. Ecclesiastes 1:18
18. Revelation 14:7 NKJV
19. Revelation 14:14 NKJV
20. Romans 6:16 NIV
21. Philippians 4:13 NKJV

22. 2 Peter 3:9
23. Rev 22:11 NKJV
24. Revelation 14:14-20
25. 1 Thessalonians 4:16, 17 NKJV
26. Daniel 12:1, 2
27. Jeremiah 31:3 NKJV
28. Matthew 4:9 paraphrased
29. Psalms 19:1 NKJV
30. Romans 1:20 NKJV
31. Genesis 1:26-28 NKJV
32. 2 Corinthians 5:17 NKJV
33. Romans 7:15-25
34. Philippians 1:6 NKJV
35. Deuteronomy 30:11-18
36. Revelation 13:8
37. 2 Thessalonians 2:8
38. Revelation 20:3 NKJV
39. John 5:28, 29 NKJV
40. Revelation 20:9 NKJV
41. Revelation 21:2 NKJV
42. Revelation 20:9b NKJV
43. Revelation 20:10 NKJV
44. Revelation 20:14 NKJV
45. 2 Peter 3:19 NKJV
46. Nahum 1:9 NKJV
47. 1 Peter 3:13
48. Rev 22:5 NKJV
49. Matthew 22:30
50. Isaiah 66:22, 23 NKJV
51. Revelation 10:5, 6 paraphrased
52. John 14:1-3 NKJV

CREATION RESOURCES

An early and personal experience with Creation will help families build strong commitment to the God of Creation. Plan your travel or vacation to visit these sites. Those who have agreed to be listed make no endorsement of this book or its contents. Dr. Wagner is happy to provide the list, but is not endorsing the sites.

The list is made alphabetically by state.

ALABAMA
DeSoto Caverns Park
Childersburg, Alabama
DesotoCavernsPark.com

ARKANSAS
The Great Passion Play
Eureka Springs Arkansas
GreatPassionPlay.org

CALIFORNIA
Adventure Safaris Exploratorium
Santa Maria, California
AdventureSafaris.org

FLORIDA
Creation Adventures Museum
Arcadia, Florida
CreationAdventuresMuseum.org

Creation Studies Institute
Ft. Lauderdale, FL
CreationStudies.org

KENTUCKY
Ark Encounter
Williamsburg, Kentucky
AnswersInGenesis.org

Creation Museum
Petersburg, Kentucky
AnswersInGenesis.org

MONTANA
Glendive Dinosaur/Fossil Museum
Glendive, Montana
CreationTruth.org

NEBRASKA
Boneyard Creation Museum
Broken Bow, Nebraska
BoneYardCreationMuseum.org

Semisaurus
Juniata, Nebraska
CreationInstruction.org

NEW YORK
Creation Museum of Upstate NY
Warrenburg, New York
henslerarc@hotmail.com

Lost World Museum
Phoenix, New York
LostWorldMuseum@gmail.com

NORTH CAROLINA
Creation Family Ministries
 Hickory, North Carolina
CreationFamilyMinistries.org

OKLAHOMA
Creation Truth Foundation, Inc.
Noble, Oklahoma
creationtruth.org

TENNESSEE
Wonders of Creation Center
Lewisburg, Tennessee
davidrives.com

TEXAS
Institute for Creation Research
Dallas, Texas
IRC.org

The Discovery Center
Abilene, Texas
evidences.org

Mt. Blanco Fossil Museum
Crosbyton, Texas
mtblanco.com

WASHINGTON
Northwest Treasures
Bothell, Washington
northwestrockandfossil.com

WISCONSIN
Living Waters Bible Camp
Westby, Wisconsin
LWBC.org

To schedule Dr. Wagner for
speaking engagements, visit
CreationRevealed.org

CPSIA information can be obtained
at www.ICGtesting.com
Printed in the USA
LVHW030753211220
674729LV00005B/168

9 781732 080607